U0323997

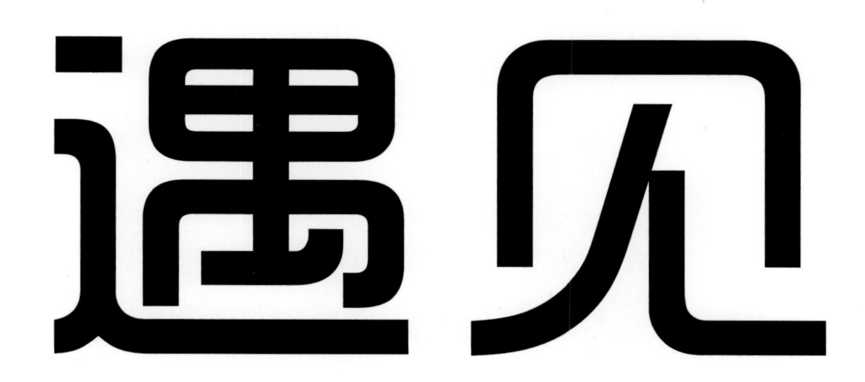

遇见

——地域特色楼盘

Rendezvousing

Characteristic Regional Property

金 盘 地 产 传 媒 有 限 公 司　策划
广州市唐艺文化传播有限公司　编著

中国林业出版社
China Forestry Publishing House

图书在版编目（CIP）数据

遇见：地域特色楼盘 / 广州市唐艺文化传播有限公司编著. -- 北京：中国林业出版社，2015.6
ISBN 978-7-5038-7988-3

Ⅰ. ①遇… Ⅱ. ①广… Ⅲ. ①建筑设计－中国－现代－图集 Ⅳ. ① TU206

中国版本图书馆 CIP 数据核字 (2015) 第 107711 号

遇见——地域特色楼盘

编　　著　广州市唐艺文化传播有限公司
责任编辑　纪　亮　王思源
策划指导　高雪梅
流程指导　黄　姗
文字编辑　殷立荣
英文编辑　冯亭亭
装帧设计　徐丹丹

出版发行　中国林业出版社
出版社地址　北京西城区德内大街刘海胡同 7 号，邮编：100009
出版社网址　http://lycb.forestry.gov.cn/
经　　销　全国新华书店
印　　刷　利丰雅高印刷（深圳）有限公司

开　　本　245 mm×325 mm
印　　张　23.25
版　　次　2015 年 6 月第 1 版
印　　次　2015 年 6 月第 1 次印刷

标准书号　ISBN 978-7-5038-7988-3
定　　价　398.00 元（USD 68.00）（精）

图书如有印装质量问题，可随时向印刷厂调换（电话：0755-86063975）。

前言 PREFACE

梁思成先生曾这样说过："建筑师的业是什么？直接地说是建筑物之创造，为社会解决衣食住行三者中的住的问题，间接地说，是文化的记录者，是历史之反照镜。"

可见建筑师的责任之重大，而在过去20多年的发展中，受经济文化发展等多因素的影响，中国的城市建设发生了翻天覆地的变化，但同时一些具有当地性格的建筑也遭受到了摧残，不符合城市特点的新建筑蜂拥而起，逐渐演变成现在千城一面的现象。回首过去的发展，我们在感叹城市发展之快速的同时，亦感受到了当地文化的遗失。

梁思成先生曾发出这样的感叹："一个东方老国的城市，在建筑上，如果完全失掉自己的艺术特性，在文化表现及观瞻方面都是大可痛心的。因这事实明显的代表着我们文化衰落，至于消灭的现象。"虽有这样的警醒在不断地提示着我们，但在过去的一段时间，受外来文化以及中国人对国外高品质生活的追求，一批又一批的西式建筑逐渐在中国城市建设中抢占到自己的领地，将西式建筑照搬照抄到中国，虽使中国人不用出国就可以感受到异域的建筑风情，但也遗忘了中国的建筑文化的根基，忘记了城市在发展过程中所遗留下来的历史印记。

一个城市的发展，都需要有自己的性格，有自己独有的特征，有自己的历史和文化。随着全球一体化的发展大潮，西式文化涌入中国，在建筑设计上，西式文化的植入，为城市增添异国色彩，但同时亦可能带来城市文化的遗失。西式文化的植入有很多种方式，其中之一是照搬照抄，这种方式弊大于利，虽带来异域风情，但也容易遗失中国建筑文化。另一个方式是中西文化的交融，以中国建筑文化文根基，以城市的历史文化为源头，只有这样，才能更有利于中国城市建设和中国建筑文化的发展。

伴随着中国房地产行业的发展及新城镇化发展的新要求，对具有地域特征和城市性格的建筑的呼声越来越高，无论是城市领导者，还是地产开发商、建筑设计师和居住者，对中国建筑文化的关注也越来越多，中国建筑文化的核心即天人合一，追求人与建筑、自然的和谐统一，这是居住者的至高追求。而中国的城市需要有中国的特色，需要有当地的特征，走在这个城市，能够领略到这个城市的历史文化及性格。

因此，基于中国建筑文化的深厚根基及发展需求以及新城镇化建设的发展需要，如何构建既符合城市发展需要，又能够传承当地文化、体验当地特色的建筑，成为城市发展的重点之一。在此，我们以地域特色楼盘为基点，精选了广东、广西、海南、江苏、上海、江西、安徽、四川、重庆、云南、西藏、北京、天津、山东等多个省市地区的30多个楼盘项目，从城市文化、地域特点等多个方面追寻中国建筑文化的根，促进中国建筑文化的传承与发展，并为城镇化建设效力。

Great architect Liang Sicheng said "What dose an architect do? From direct meaning, it is to create buildings to solve one of the three living requisites in our life; while from another meaning, it is a recorder tracking culture and a mirror reflecting history."

From the words of Liang, we know that architects bear significant responsibilities. However, in the past 20 years in China, city construction has experienced dramatic changes due to economical and cultural developments. Great amounts of vernacular architecture has been damaged, while a good many of new buildings have been springing up, which made lots of cities appear similar faces in China. In retrospect, we have to regret for regional culture loss when we exclaim over our cities' rapid developments.

Liang Sicheng had sighed like this "It is a pity that an ancient oriental country loses its typical arts on architecture not only on cultural manifestation or on appearance appreciation, because this means that our culture is declining or disappearing gradually." Although this kind of warnings have reminded us regularly, we still underwent a foreign cultural impacts. Because Chinese people yearn for foreign quality life, we introduced in groups of western architecture, and they are gradually occupy Chinese cities. Western architecture copy tide makes Chinese people feel exotic architectural flavors at home, while we start forgetting our own architectural and cultural roots and those historic stamps leaving in the process of city developments.

The growth of every city cannot discard its unique dispositions, characters, history and cultures. With the development of the globalization tide, western culture has swarmed into China. Western architecture transplant adds an exotic flavor to a city, while it changes the city's cultural developing trajectory at the same time. Actually, foreign culture transplant has more than one way: copy or integration. Copy will damage Chinese architecture cultural developments, while integration will promote it. In the process of Chinese-foreign cultural integration, we should hold out cultural foundations and trace back to historic sources. Only in this way can we build better cities and boost Chinese architectural culture.

With developments of Chinese real estate industry and new urbanization requirements, governmental leaders, real estate developers, architectural designers and inhabitants have concerned Chinese architectural cultures, and supporting has been growing for regional and urban characteristic buildings. Chinese architectural culture focuses on unity of human and heaven, and it is a pursuit of supreme that human, architecture and nature are in a perfect harmony. Chinese cities ask for Chinese features and vernacular characteristics so that we can feel the cities' historical and cultural spirits when we visit them.

Basing on profound cultural foundations and requirements of new urbanization, how to build characteristic regional architecture to echo city's developments and inherit vernacular cultures attracts wide attention in China. Here, we selected over 30 projects from Guangdong, Guangxi, Hainan, Jiangsu, Shanghai, Jiangxi, Anhui, Sichuan, Chongqing, Yunnan, Tibet, Shanxi, Beijing, Tianjin and Shandong to present Chinese architecture cultural roots from historical and regional culture aspects. We hope this can make a contributing to Chinese architecture cultural inheritance and developments and render a service to urbanization construction.

目录 CONTENTS

沪苏诗意

Poetic Shanghai and Jiangsu

徽州印象

Huizhou Impression

京鲁风韵

The Charm of Beijing and Shandong

目录 CONTENTS

川渝民居

Sichuan-Chongqing Dwellings

岭南画意

Lingnan Scroll

滇藏如画

Picturesque Yunnan-Tibet

沪苏诗意

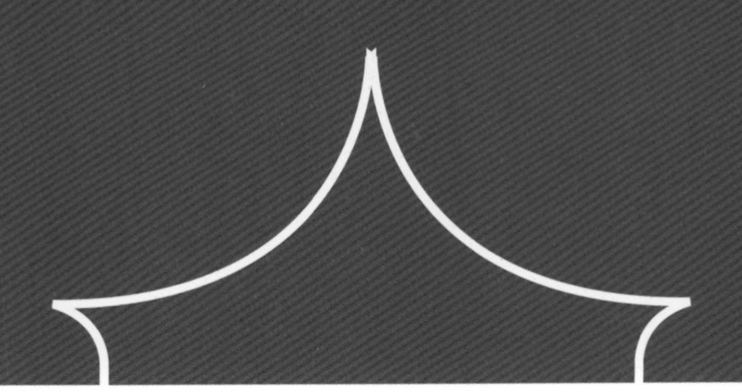

Poetic Shanghai and Jiangsu

沪苏区域涵盖上海、江苏、浙江三个地区。之所以将这三个地区划分为同一类，源于其建筑文化的根源，即富有诗意的江南园林风格。沪苏建筑有着小桥流水的诗情画意，亦有着历史厚重。

苏州园林的知名，古镇文化的影响，将江南园林建筑集中于江苏一带。上海建筑以石库门著名，石库门就是上海的一张名片，但追根溯源，石库门却是脱胎于江南民居的住宅形式。石库门建筑保持着中国传统建筑以中轴线左右对称布局的特点。而随着历史的发展变迁，受外来文化的影响，石库门逐渐演变为融合中西建筑艺术的建筑和文化的产物，不仅有着江南传统民居的空间组织，还融合了西方的建筑元素和文化。而在历史的发展变革中除了石库门带有浓厚的中西合璧的色彩外，20世纪初出现的民国建筑也融合了西方的建筑元素，而南京的民国建筑更多地体现出江南园林的意蕴和格调。

在此，我们选定浙江、江苏和上海地区的一些项目，根据当地经济、文化的发展及自然地理、气候和生活习惯等因素的变化，从民族性和现代性两个方面来阐述项目的场所精神和地域性，在保留传统文化的同时，利用现代的科学技术及人文背景传承和发扬江南园林建筑。

历史的印记随着时代的发展也在不断地变迁，这里有富有民国特色的建筑，有对原有建筑的还原与保护，也有加入现代元素的传承，杭州湖边邨酒店保留了原有风貌，南京颐和公馆酒店则活化再利用了原有的民国建筑，它们更多的呈现了历史性、传统性和民族性，而南京宏图水上庭院虽采用了民国风，但加入了更多的现代元素，所呈现的更多是现代性。

如何更为纯粹和自然地传达出江南园林风格是值得深思的议题。苏州桃花源、朱家角九间堂西苑关注江南民居的院落感和聚落感及街巷文化，昆山锦溪岛尚、扬州虹桥坊则是从江南水乡的多元文化及历史人文中挖掘题材，突破设计。在追求传统的同时融合现代元素，注重细节的创新，呈现出现代性与当代性。

此外，我们还选择了具有元素移植和复制特点的项目，如具有石库门特点的长春绿地中央广场•饕界和具有江南水乡特色的天津中国国家画院盘龙谷创作基地，与前面的项目形成对比，突出设计的地域性和场所精神。

This chapter includes two regions, Shanghai and Zhejiang, because they have common origin of architectural culture, that is poetic Jiangnan garden style. Shanghai and Jiangsu buildings make people associate with poetic environment of bridges with flowing streams, whereas they also have heavy history.

Famous Suzhou garden and ancient town culture converge Jiangnan gardens in Jiangsu. Shikumen architecture is a name card standing for Shanghai. Tracing back to origins, Shikumen was evolved from Jiangnan folk residence style, and until now it remains a Chinese traditional architecture feature: symmetric central axis distribution. With progress of history and influence from foreign culture, Shikumen has developed to an architectural and cultural product blending Chinese and foreign architectural art. It contains traditional Jiangnan dwelling spatial organization and western architectural elements. Republican period architecture in the early 20th century intermingled western architectural elements, and at that time Nanjing architecture suffused Jiangnan garden style.

Projects are collected from Zhejiang and Shanghai in the chapter. According to local economy, culture development, natural geography, temperature and life style, we analyze these projects' place spirits and regional characteristics from perspectives of national character and modernity. We exert to maintain tradition, meanwhile, take advantage of modern scientific techniques and cultural context to inherit and develop the Jiangnan gardens.

Historic stamps are developing with times processing. Here we can enjoy republican period featured architecture, conservation and restoration for original buildings and traditional buildings appending modern elements. Hangzhou Relais & Chateaux persists the original appearance; Najing Yihe Mansions revivify republican period architecture; both of them present historic significance, traditional characteristics and national characters. Nanjing Gentler River Villa adopts republican period style while adds modern elements, so it takes on more modern features.

There is another four projects stand for Jiangnan garden style. Suzhou Taohuayuan Villa and Shanghai Zhujiajiao Nine-Row Mansion express Jiangnan dwelling courtyard settlement and alley system cultures; Kunshan Jinxi Dao Shang Villa and Yangzhou Rainbow Square excavate multicultural and historical humanism elements from Jiangnan water towns, which is a dramatic breakthrough in design. Pursuing tradition while intermingling with modern elements, the design conceptions take detail innovation to foreground, featuring in modern and contemporary features.

In addition, we chose two projects with elements transplanting and duplicated features, such as Changchun Greenland Central Plaza Appetiting Dragon featuring in Shikumen style and Panlong Valley National Painting and Calligraphy Creation Base boasting Jiangnan water town flavor. The two projects highlight regionalism and place spirits comparing to the mentioned projects above.

传统聚落空间 江南古镇风情

上海朱家角九间堂西苑

Traditional Settlement Space Jiangnan Ancient Town Style

Zhujiajiao Nine-Row Mansion, Shanghai

开发商：证大三角洲置业有限公司　建筑设计：加拿大CPC建筑设计顾问有限公司　项目地址：上海青浦区朱家角古镇西北角
联排用地面积：24 197平方米　联排建筑面积：21 100平方米　大宅用地面积：7 360平方米　大宅建筑面积：4 400平方米　采编：康欣

Developer: Zendai Group Sanjiaozhou Real Estate Co., Ltd.　Architecture Design: The C.P.C. Group
Location: Northwest of Zhujiajiao Ancient Township, Qingpu District, Shanghai
Townhouse Site Area: 24,197 m²　Townhouse Building Area: 21,100 m²　Mansion Site Area: 7,360 m²　Mansion Building Area: 4,400 m²
Contributing Coordinator: Kang Xin

孕育于江南，根植于朱家角，朱家角九间堂西苑通过天然河水的引入、江南水乡村落布局与朱家角民居院落空间的营造将江南水乡情怀表达得淋漓尽致。月洞门、影壁、山石、雕刻等元素的运用，演绎出中国传统江南古典园林的韵味，透露出一份艺术、雅致、古朴的气息。在传承传统的同时，对于新材料、新技术的运用，又将古典与现代相交融，在传承传统的同时散发出现代的气息。

Breeding in Jiangnan region and rooting in Zhujiajiao, the Zhujiajiao Nine-Row Mansion creates a incisive and vivid Jiangnan water town through introducing natural rivers, Jiangnan village layout and Zhujiajiao dwelling courtyard spaces. Chinese traditional Jiangnan classical garden elements: moon gate, screen wall, rockwork and sculpt emit out an artistic, elegant and pristine breath. Many advanced materials and technologies are applied in the project to blend in classic aesthetics, so the project boasts an integration of classic and modern.

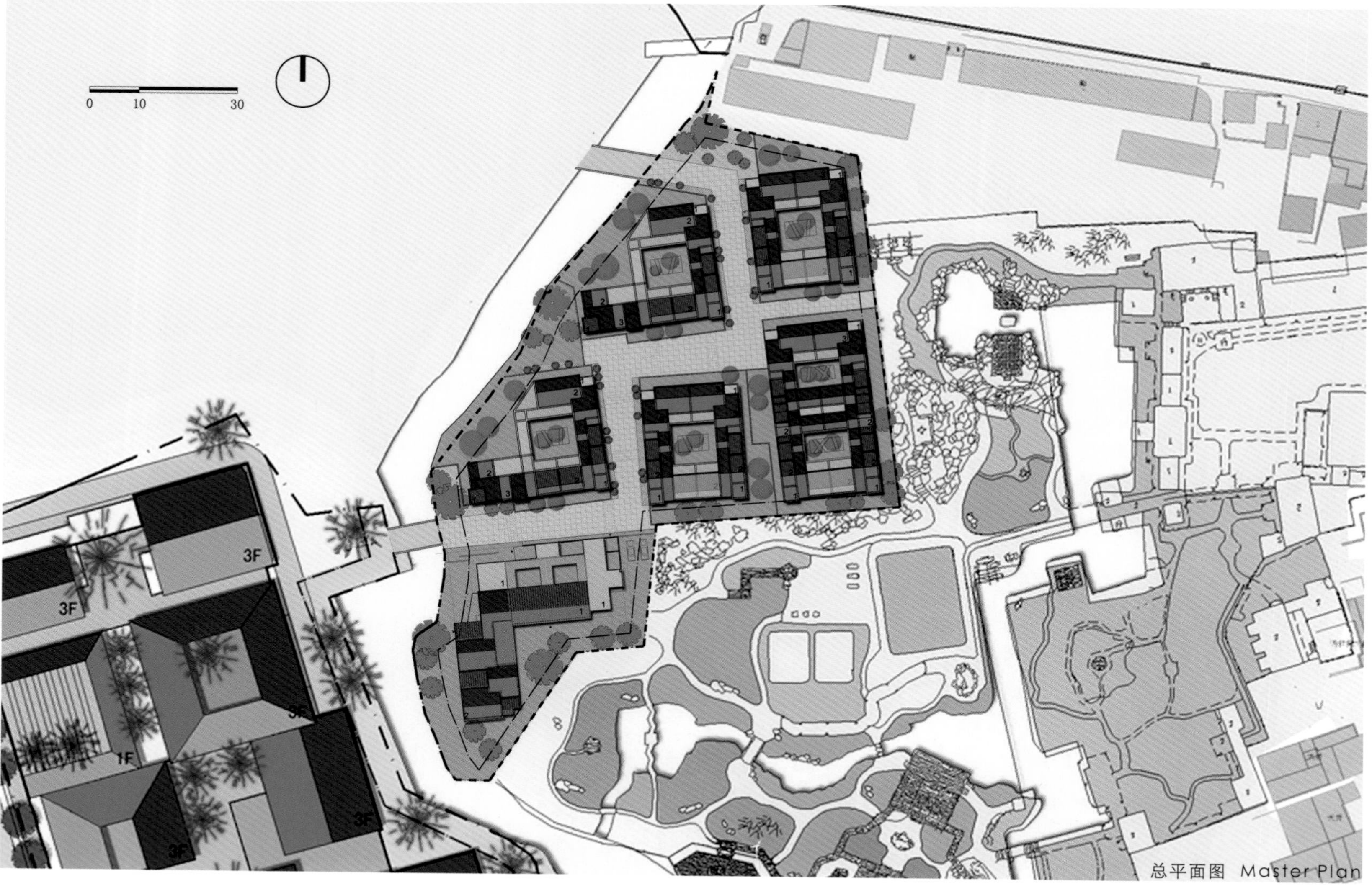

总平面图 Master Plan

西立面图 West Elevation

北立面图 North Elevation

东立面图 East Elevation

南立面图 South Elevation

9.100
6.600 3F
3.600 2F
±0.000 1F
-0.150
2500
3000
3600
9250
150

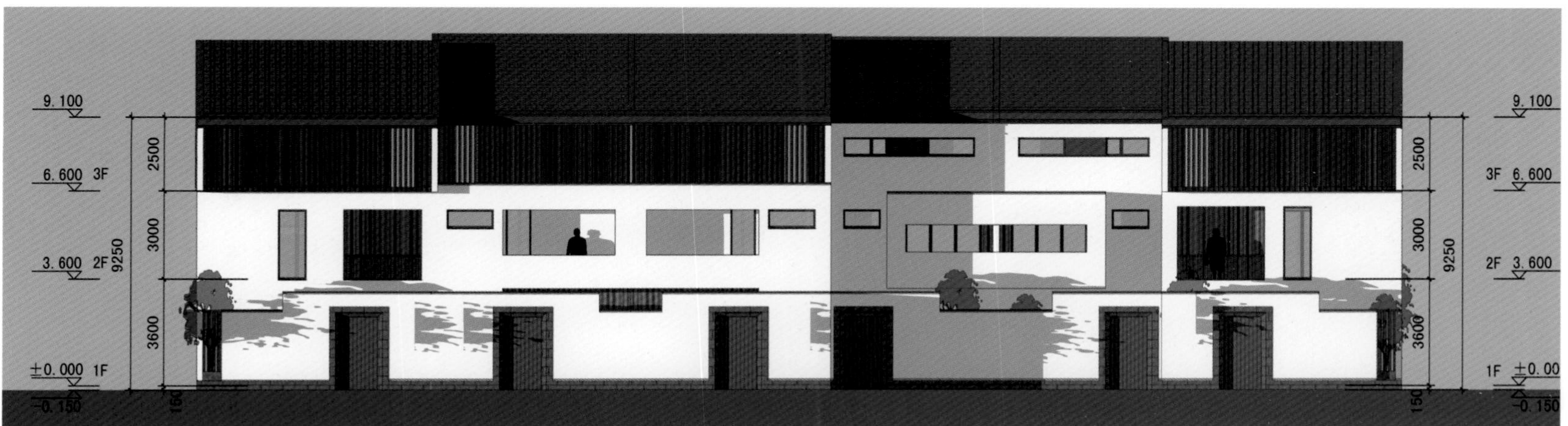
9.100
6.600 3F
3.600 2F
±0.000 1F
-0.150
2500
3000
3600
9250
150

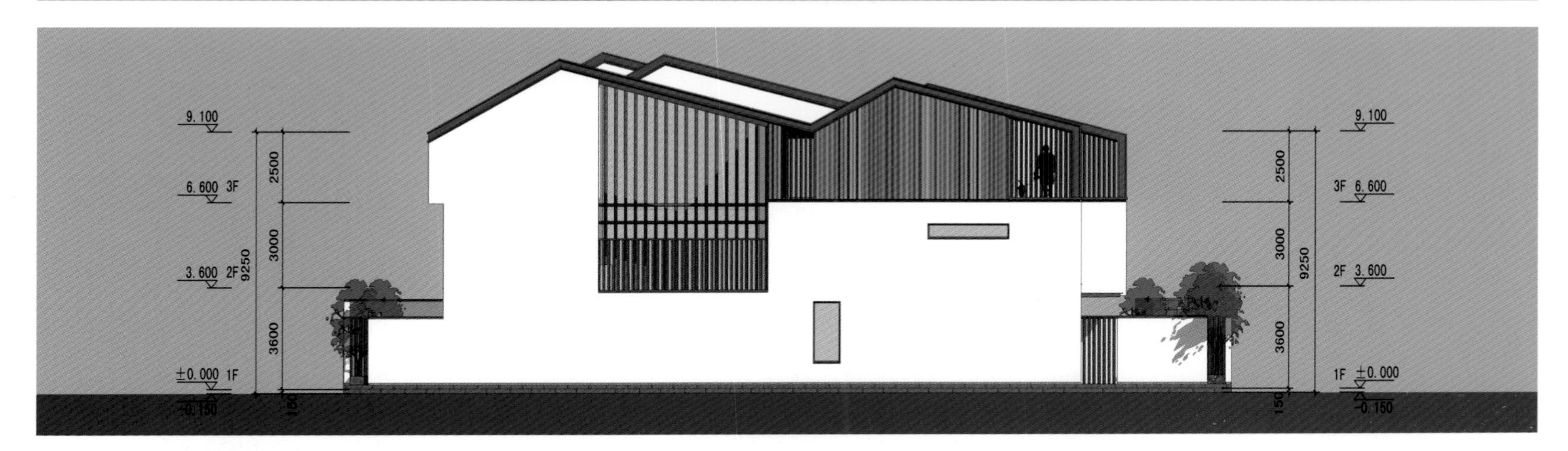
9.100
6.600 3F
3.600 2F
±0.000 1F
-0.150
2500
3000
3600
9250
150

区位概况

项目北靠大淀湖，紧邻朱家角古镇风景区，东接课植园，环境优美，拥有超高绿化率，绝佳的自然条件造就了宜居的生态社区。随着附近虹桥交通枢纽、淀山湖大道、轨道17号线畅达市中心，该地段的区域价值拥有巨大的升值空间。

规划：水乡古镇

项目在水乡古镇朱家角的保护控制范围内，总体规划指导思想延续朱家角江南古镇的特色肌理和建筑文脉，将新的生活方式和传统相融合，发扬和传承江南水乡的建筑传统。

项目以“自然和谐，生态宜居”为目标，营造新中式低密度居住环境，以江南传统村落的鱼骨状结构布局，以朱家角民居院落为空间特色，营造出空间层次丰富，建筑风格鲜明的院落别墅群。

Location Overview

The project is against the Dadian Lake in the north and close to Zhujiajiao Ancient Township scenic zone. It connects to Kezhi Garden in the east, which enjoys beautiful environment and high greening rate. Usually extraordinary natural conditions create a livable community, and Hongqiao transportation hub, Dianshanhu Avenue and Rail line 17 offer convenient transportation to downtown, so the region has huge appreciation potential.

Planning: Waterside Ancient Town

The project is in the conservation area of Zhujiajiao Ancient Township. In order to carry forward architectural tradition of Jiangnan water town, the general planning adopts Zhujiajiao-featured texture and Jiangnan context to merge new life-style into tradition.

Aiming to build a natural, harmonious and livable community with a neo-Chinese low-density living environment, herringbone texture is employed in the layout. Taking Zhujiajiao dwelling courtyard as space features, the design builds villa groups with multileveled spaces and distinctive styles.

小贴士

朱家角作为千年古镇闻名遐迩，镇内河港纵横，青砖黛瓦的明清建筑依水而立，蜿蜒曲折的小巷、花岗岩石的街面、古风犹存的36座石桥形成古朴自然的风光，其小桥、流水、人家的格局和水乡古镇的风情吸引了无数的文人骚客，也让很多游客流连忘返。

Tips

Zhujiajiao Ancient Township is a millennium eldest town gaining widespread renown. There is a grid of rivers and harbors inserting in the town where exist buildings from Ming and Qing dynasties cladding blue bricks and black roof-tiles standing by rivers, twists and turns lanes with granite paving, and 36 pristine antique arch stone bridges. Waterside houses and amorous feelings of the waterfront town fascinate millions of visitors.

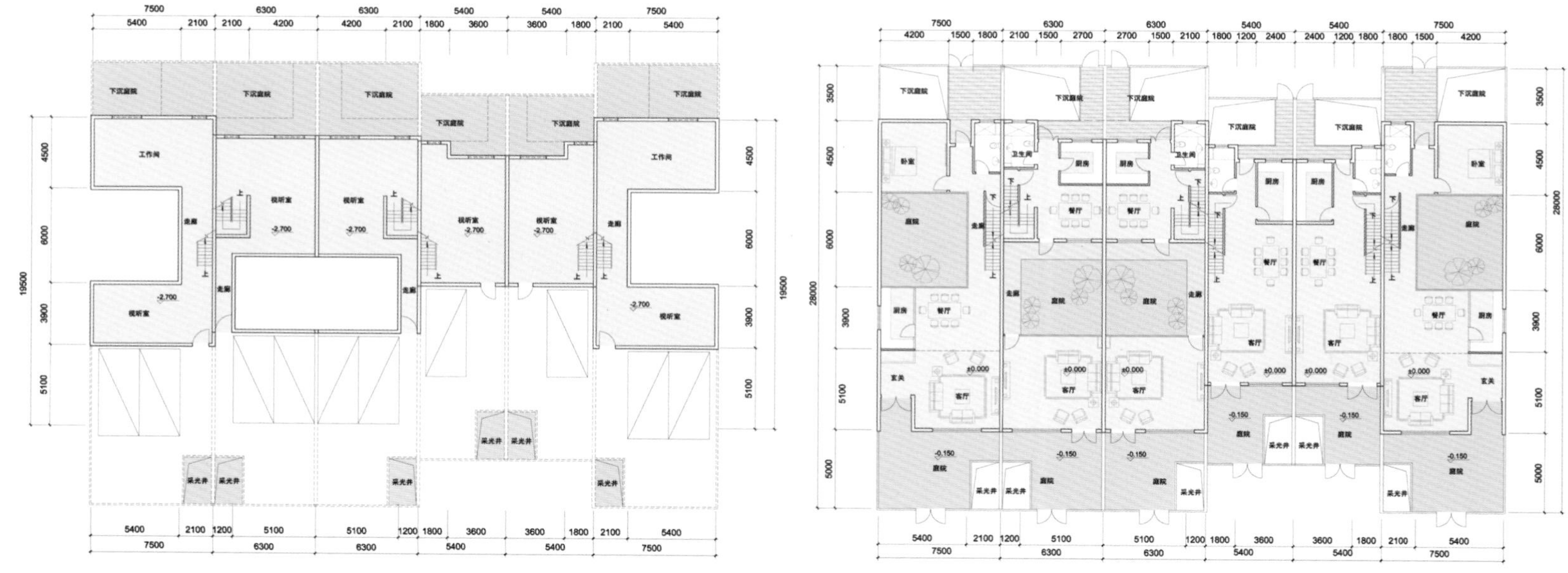

组团地下一层平面 Basement Plan

组团一层平面 1F Plan

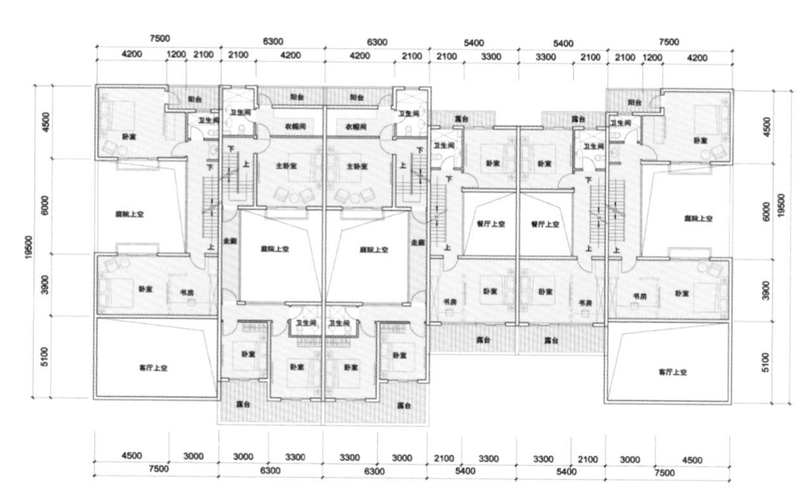

组团二层平面 2F Plan

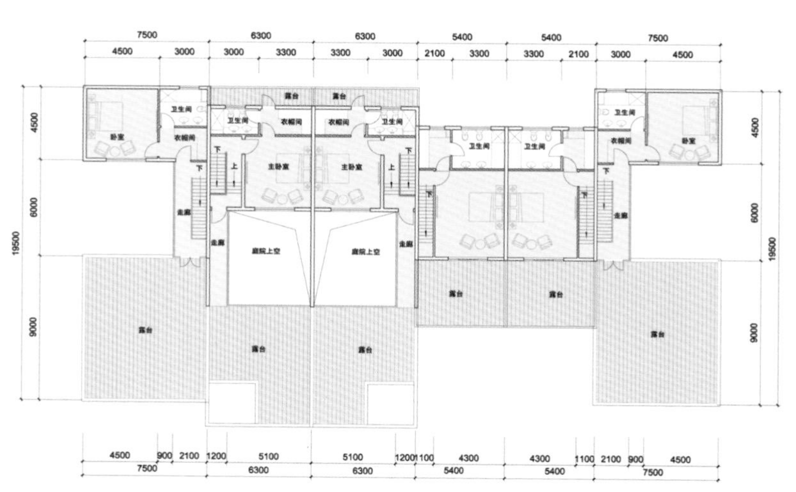

组团三层平面 3F Plan

建筑设计：创新演绎

设计师强调通过对比例、形体关系、色彩和材料等的控制，实现对传统建筑构造和建造方式的研究和演绎，在立面形态上营造出传统中式建筑的意向和联想，同时避免过多的符号和复杂手法。设计还积极探索使用新技术和新材料，以期用现代的方式演绎和传承传统建筑文化。

景观设计：古典元素

景观设计将天然河水引入社区，强调江南水乡小镇的规划特点，演绎出中国江南水乡风格的古典园林精神，塑造丰富、变幻、生动而醇美的空间与景致。通过月洞门、影壁、山石、雕刻等随处可见的古典园林元素，让整体氛围透着雅致的古典艺术品味。

户型设计

联排面积约195～400平方米，独门独户设计，多数房型设有多露台，便于随心安排不同功能。部分房型附赠约100平方米超大地下室，使得可利用空间更为阔绰。户型设计在南北进深方向分别结合建筑的使用功能布置了前院、中庭、后院三个院落空间，形成秩序鲜明的空间递进关系。

Architectural Design: Innovation

Only lay great emphasis on study of controlling proportion, configuration relations, colors and materials can designers perform a sucessful traditional building structure and a construction mode for the project. Chinese traditional architectural symbols are simplified on facade while prevent excess expression and sophisticated design methods. Designers positively explore state-of-the-art technologies and advanced materials so that traditional architectural culture can be demonstrated and passed on in a modern way.

Landscape Design: Classical Element

Leading natural river water into the community and highlighting water town planning features successfully display a classical garden spirit of Jiangnan water town. Plentiful, various and dynamic spaces and views reflect on ubiquitous classical garden details, such as moon gate, screen wall, rockwork and sculpt, making the overall ambiance full of elegance and classical artistic tastes.

House Type Design

The townhouse area is 195 m² to 400 m², and every unit enjoys independent entrance. Most of house types equip with terrace for residents setting various functions at their will. Some house types are complimentary with about 100 m² super-sized basement to expand available spaces. Front yard, atrium and rear yard are in north-south orientation, forming a sequential and progressive space relation.

海派院落空间 演绎江南意境

上海金茂崇明凯悦酒店

Shanghai Style Courtyard Space Display Jiangnan Mood

Chongming Hyatt Regency, Shanghai

开发商：金茂（上海）置业有限公司　建筑设计：上海骏地建筑设计咨询有限公司（JWDA）　设计师：郑士寿、潘嵘
景观设计：贝尔高林国际（香港）有限公司　室内设计：WILSON　项目地址：上海崇明县陈家镇　用地面积：34 000平方米　建筑面积：48 000平方米

Developer: JinMao Property Group (Shanghai)　Architecture Design: JWDA　Designers: Zheng Shishou　Pan Rong　Landscape Design: Belt Collins
Interior Design: WILSON　Location: Chenjia Town, Chongming County, Shanghai　Site Area: 34,000 m²　Building Area: 48,000 m²

海派建筑根源于江南园林，其对自然的运用、对院落的领悟都遵循着江南所独有的韵味，呈现的是一片富有诗意的空间。同时，作为时尚前沿城市，又有着海纳百川的特性。上海金茂崇明凯悦酒店设计呈现出传统与现代结合的美感。设计与周边的自然景观相结合，以院落为串联主线的空间模式，将木架结构和现代混凝土结构相互穿插，将传统的中式元素与现代技术相融合。而细节上的坡顶、山墙和木窗格以及桥、廊、亭、台等元素的运用，营造出富有诗意的意境美，完美地展示了上海本土的江南意境空间。

Shanghai style architecture can trace back to Jiangnan garden, which features in using nature and complying to courtyard layout, so it naturally presents a poetic space. Shanghai is a fashion-forward city, boasting inclusive characteristics. The project expresses a sense of beauty, subtly combining tradition and modernism. The design well integrates with peripheral natural landscapes. The spatial model is tandem courtyards; timber frame and modern concrete structure interweave and invite traditional elements and modern technologies. Slope roof, gable wall, wooden window, bridge, colonnade, pavilion and terrace bring poetic beauty and demonstrate a Shanghai local Jiangnan space.

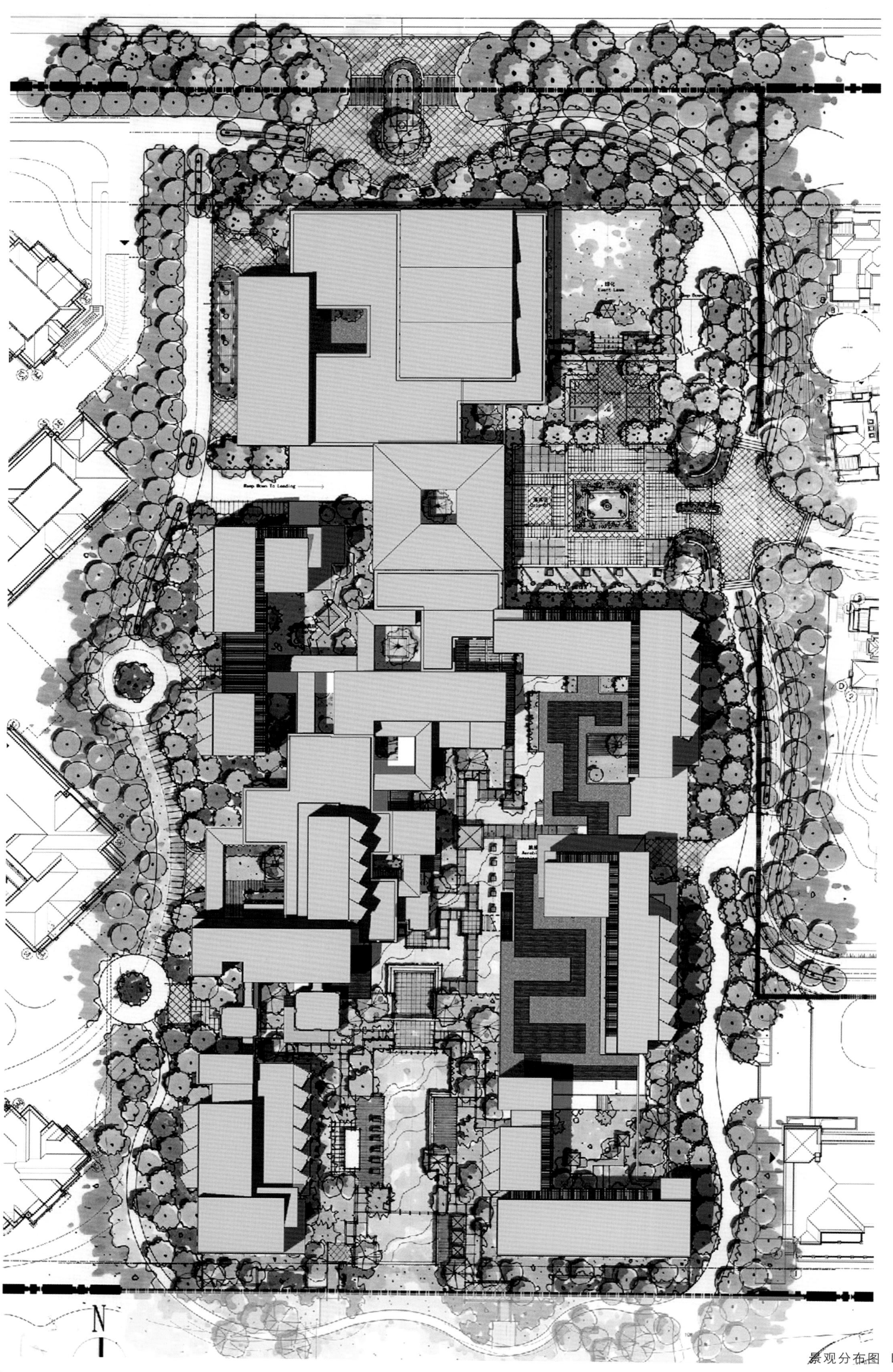

景观分布图 Landscape Plan

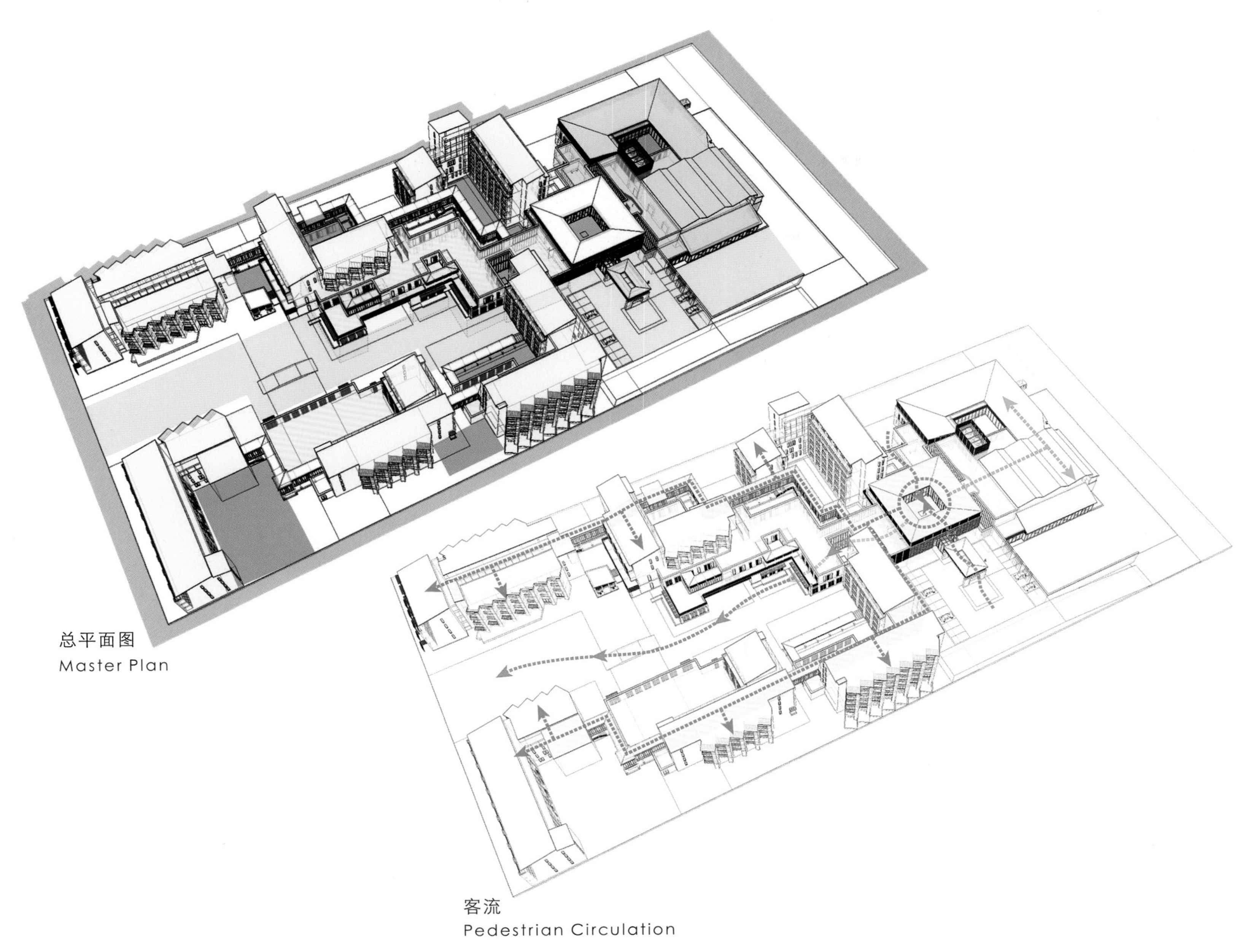

总平面图
Master Plan

客流
Pedestrian Circulation

区位分析

项目坐落于被誉为“长江门户，东海瀛洲”的崇明岛，邻近我国南北海岸线上最大的自然保护区——崇明东滩鸟湿地公园。基地东至规划纵三河，西至规划纵二河，南至规划中心河，北至揽海路。地块呈微微弯曲的四边形，地势较为平坦。

规划理念

项目设计没有遵循星级酒店一贯的气派、高效、震撼的做法，而是采用一连串分散的、低矮的、以院落为串联主线的空间模式，用大小不同、形状不同的十几个院子将酒店的每个功能空间连接在一起，园园相接，层层递进，并保证与周边环境相融，鸟类的安宁生活也不至于被人类的过度开发而打扰。

规划布局

木结构的四合院大堂四面对称，中心稳定，富有禅意。大堂向北是宴会厅、培训中心和娱乐中心；大堂南面布置有一东一西两条木结构廊桥，它们各自连接起3栋独立的高低错落的客房楼；大堂向南下一层是酒店的公共空间，设有餐厅、茶楼、spa、健身房、游泳池等休闲设施。

Location Overview

The hotel locates on the Chongming Island that renowned as a Yangtze River Portal and Sanctury of East China Sea, neighboring with the Chongming Dongtan Bird Island Wetland Park which is the largest natural conservation area on south-north coastline of China. The site extends east to the Planed Zongsan River, west to the Planed Zong'er River, south to the Planed Central River, and north to the Lanhai Road. The plot appears a curved quadrangle with flat terrain.

Planning Conception

The hotel does not follow a routine design for starred hotels that pursues style, efficiency and shock, but employs a space model of a succession of scattered and low-rise courtyards concatenated into hotel spaces. Every functional space is connected by a dozen of vary-sized and variform courtyards, one-by-one, layered progressing, to ensure a compatibility with surrounding environment as well as birds' peaceful life.

Planning Layout

The timber structure quadrangle courtyard is in symmetrical layout at four sides and has a steady center, flaunting an atmosphere of Zen. Standing at principal hall, at the north is a banquet hall, a training center and an entertainment hall in sequence; at the south of the principal hall two wooden lounge bridges connect three independent different-height guest room buildings, one in the east, one in the west; public spaces are downstairs at the south of the principal hall including a Chinese restaurant, a tea house, spa rooms, a gym center, a swimming pool, etc.

小贴士

木结构是单纯由木材或主要由木材承受荷载的结构，通过各种金属连接件或榫卯手段进行连接和固定。这种结构因为是由天然材料所组成，受材料本身条件的限制。而斗栱是中国木架建结构中的关键部件，其作用是在柱子上伸出悬臂梁承托出檐部分的重量。紫禁城太和殿是现存最高等级的木构古建筑。

Tips

Timber Structure is a load-bearing framework simply made of wood or mainly of wood, and they are fixed by metal joint connectors or tenon-and-mortise technique. This kind of structure is made by natural materials, so it is inevitably restricted by its own natures. Dou-gong is a critical component of timber structure in Chinese architecture, because it is used for supporting the weight of eaves out of cantilever beams. The Taihe Palace in the Forbidden City is the existing supreme grade timber structure building.

建筑设计

项目以现代中式风格为整体格调，与周边生态环境自然融合，在时尚现代的基础上融入上海本土元素和度假风格，带有独特的审美情趣。整个建筑由代表传统的木结构和代表现代建筑的混凝土结构穿插而成，木结构的调和使得本来紧张刻板的建筑变得轻松自由，富有弹性和张力，木结构、石材和粉刷涂料结合在一起，就像江南传统的黑白灰，非常吻合传统审美。三种颜色、三种材质、三种体量，用中国的空间逻辑互相缠绕交织在一起，坡顶、山墙、木窗格隐约在其中，呈现出诗意的意境美。

景观设计

景观设计将东方园林空间的韵味与以分散休闲为特征的度假酒店有机融合，营造步移景易的体验，使建筑成为庭院的背景，与绿化交融，寻求生态环境与舒适人居的平衡发展。户外空间的尺度或大或小，或宽或窄，或直或转，通过河、桥、台、榭、石等景观元素穿插其中，让人仿佛走在水乡街巷的某个角落。

Architectural Design

The hotel uses modern Chinese style as keynote, and well fits into surrounding natural ecological environments. Basing on a modern and stylish flair, it complements with Shanghai local elements and a resort style, bringing a unique aesthetic taste. Timber structure represents tradition and concrete structure boasts modernity, and both of them compose the hotel. The timber structure produces a harmonious effect that turns original stiff building to a relatively free, easy and elastic one. The wood, stone and plaster coating combine to form a Jiangnan traditional color palette: black, white and gray, coinciding with traditional aesthetic appreciation. Three colors, three materials and three volumes interweave with one another, and slope roof, gable wall and wooden window are occasionally seen among them, which interprets a poetic beauty.

Landscape Design

The landscape design infuses an oriental courtyard flavor with resort hotel leisure, creating an effect of scenery changing with moving steps. The architecture serves as a foil to courtyards, and incorporates with greening to pursue a balance between ecological environment and comfort habitation. Outdoor spaces are of various sizes, wide or narrow, straight or twisted, and match with landscapes such as stream, bridge, stage, pavilion, rock, etc., make you expose in a corner of a water town alley.

传统院落空间 苏州园林新体验

苏州绿城桃花源

Traditional Courtyard Space New Experience in Suzhou Gardens

Taohuayuan Villa, Suzhou

开发商：苏州绿城玫瑰园房地产开发有限公司 建筑设计：浙江绿城建筑设计有限公司 项目地址：苏州工业园区独墅湖北
占地面积：213 853平方米 建筑面积：232 952平方米 容积率：0.6 绿化率：30% 采编：康欣

Developer: Suzhou Green Town Rose Garden Real Estate Development Co., Ltd. Architecture Design: GAD Location: Dushu Hubei in China-Singapore Industrial Park
Site Area: 213,853 m² Building Area: 232,952 m² Plot Ratio: 0.6 Greening Rate: 30% Contributing Coordinator: Kang Xin

江南建筑以轻巧、秀美、雅致而著称，其粉墙黛瓦的色彩，配合着迷人的水乡景色，刻画出一幅浓淡相宜的水墨画。苏州绿城桃花源将江南建筑中粉墙黛瓦的古典建筑特色与自然景观形成色彩搭配，在原汁原味地还原苏州建筑特色的同时凸显细节的精致，并将传统院落的聚落感及苏州园林和街巷文化进行了全新的演绎。通过流动的巷道空间与新中式的院落空间布局，使小环境与大环境协调统一，传达出传统园林亭台楼阁的空间意境，带领居住者体验中国传统建筑的院落感。

Jiangnan architecture is known for its lithe, beauty and elegance, which is like an inkwash painting depicting a water town landscape with white wall and black tile. Jiangnan classical architecture white walls and black roof-tiles match natural sceneries harmoniously. The project not only restores original Suzhou architecture but brings delicate details to the foreground, displaying settlement sense of traditional courtyard and alley culture. Fluid alley space and neo-Chinese courtyard layout reach a harmonization in small and broad environments, offering residents a traditional Chinese courtyard living experience.

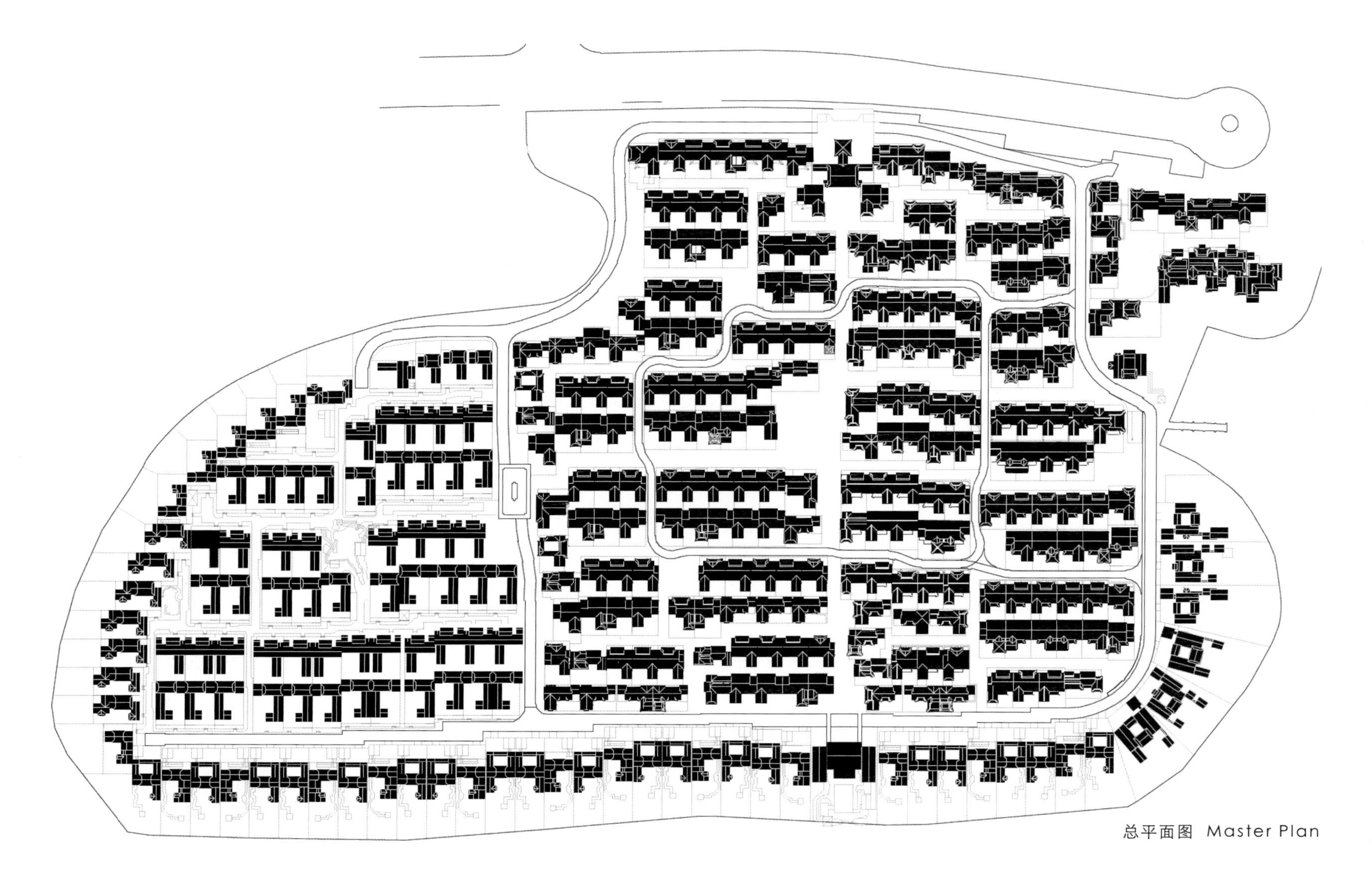

总平面图 Master Plan

区位概况

项目基地位于苏州市东部，苏州工业园区中部，金鸡湖东南侧。周边生活配套设施便利，学校、银行、医院、公园等公共服务都近在咫尺。整个地块地势平坦，三面临水，南部直面独墅湖，北侧与高尔夫花园住区隔街相邻，东侧与建设中的高尔夫球场隔水相望，西侧为大湖城邦花园。

项目背景

苏州地区河道纵横，湖荡罗列，为了适应这种多水的自然生态环境，苏州城市有其独特的水陆交通体系和民居建筑形态，并组合成具有江南特色的城市空间，达到了生态、文态的有机统一。苏州民居多临河依水，或沿河而建，形成水上小巷；或背河而建，前街后河，形成街巷，随着历史的发展，粉墙蠡窗，照影映波，形成具有苏州特色的水乡民居风貌与街巷文化。而苏州古典园林"宅园合一"的特点更是深入人心。

Location Overview

The project locates in the east of Suzhou, in the middle of China-Singapore Suzhou Industrial Park, and close southeast of the Jinji Lake. The peripheral living facilities are convenient: school, bank, hospital and park all near at hand. The plot is flat with water at three sides, and it faces the Dushu Lake in the south, overlooks Golf Garden community across a street in the north, and the golf course is under construction on the other side of the lake, and Lakeside City lies in the west.

Project Context

Crisscross rivers and scattered lakes spread in Suzhou. In order to adapt multiwater natural ecological environments, Suzhou has a unique amphibious transportation system and civilian dwelling configuration, and the design creates Jiangnan featured urban spaces and an organic unity of ecology and cultural contexts. Suzhou style civilian dwellings stand at riversides, forming water alley; or back against rivers and face streets, forming street alley. With the progress of history, white walls and shell windows, reflection of mountains forming perfect pictures in the pool, these featured scenes come into Suzhou featured water town dwellings and alley culture. Suzhou classical garden layout "mansion blending in garden" attracts myriad souls.

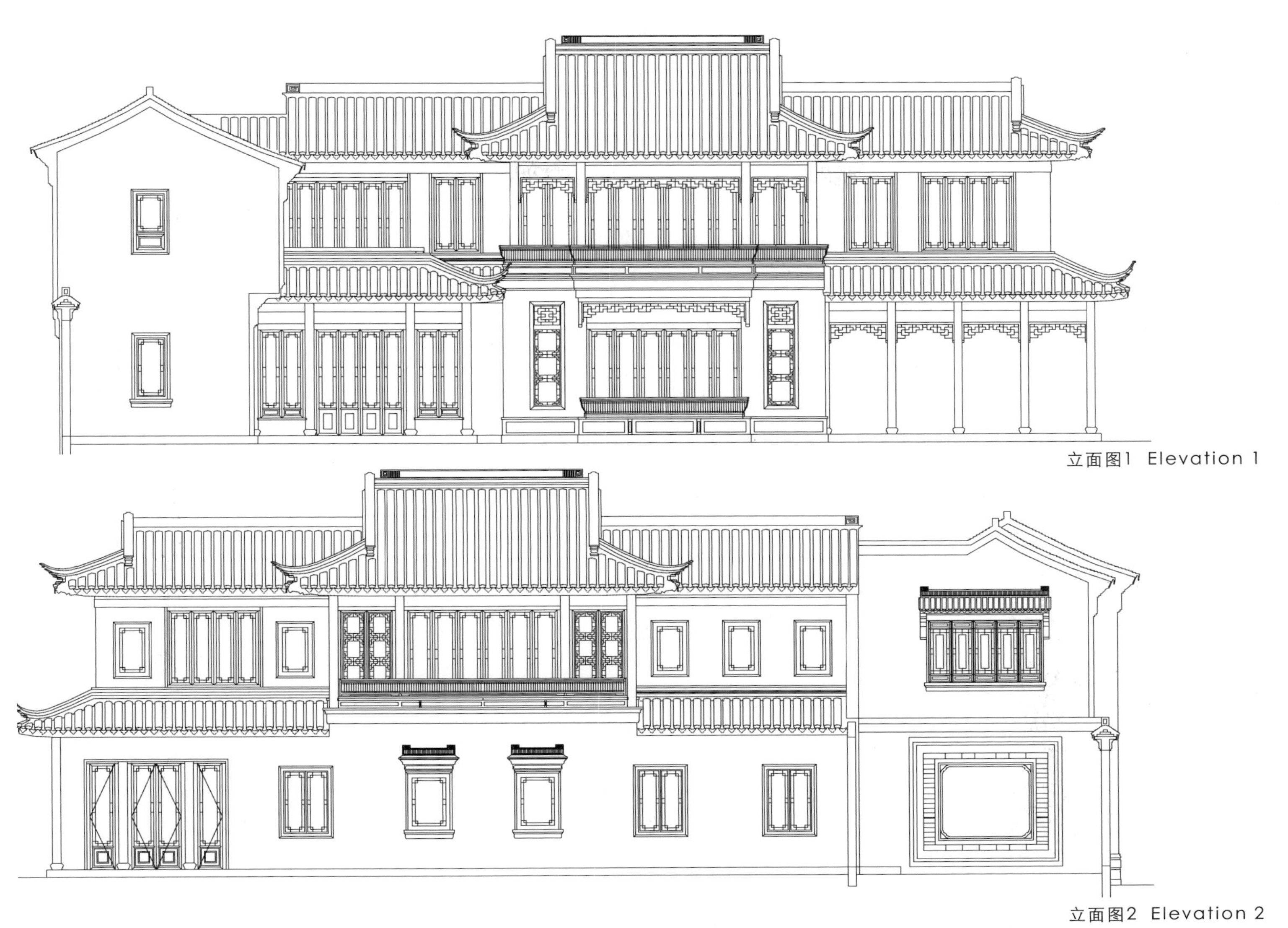

立面图1 Elevation 1

立面图2 Elevation 2

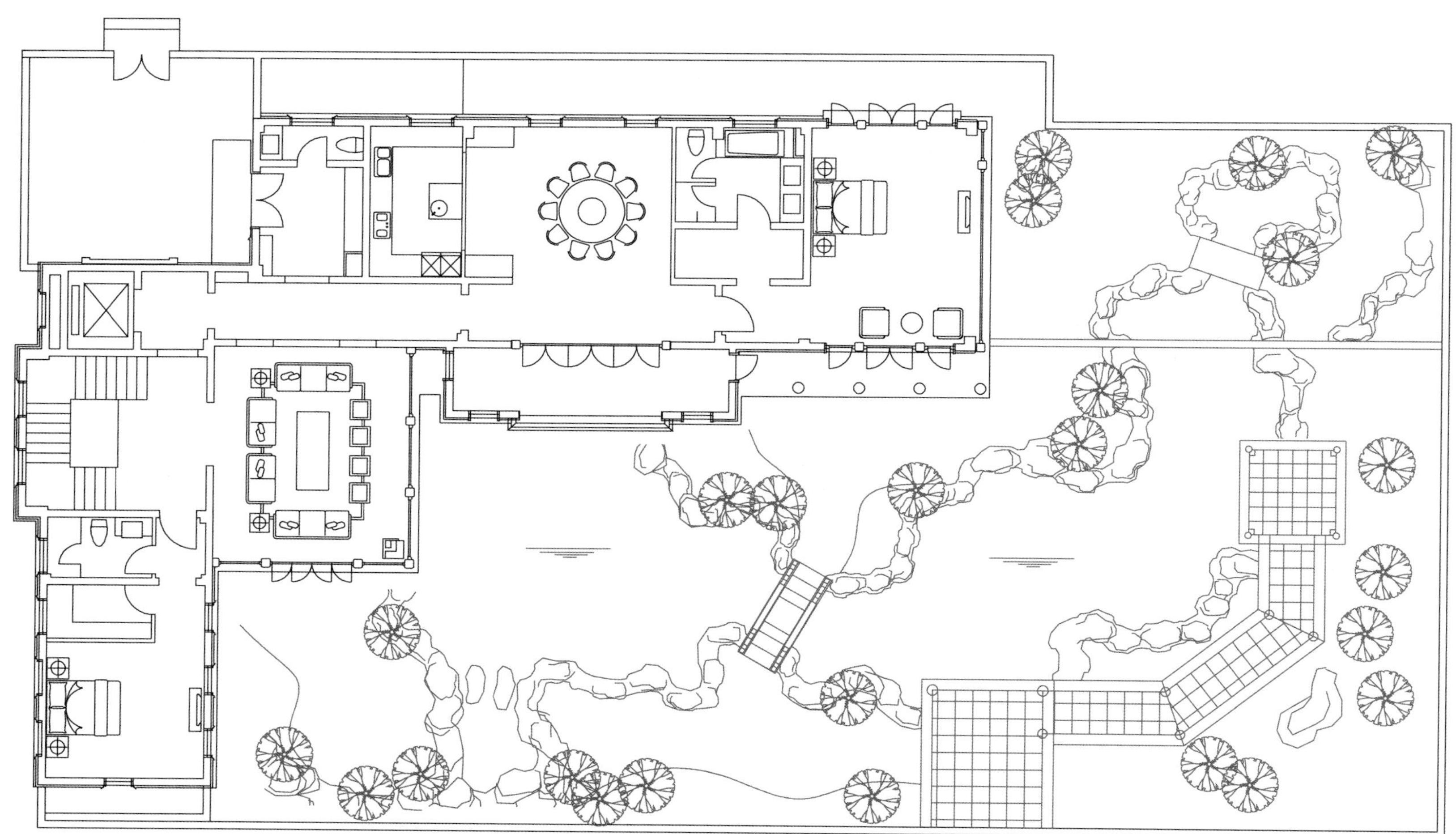

首层平面图 1F Plan

规划设计

整体住宅布局分为以合院住宅为核心的两大片区：一是基地中部的合院住宅，分为东中西三区，依托错位曲折的双十字空间轴线纵横展开并相互关联，营造层次丰富，意象鲜明的社区核心空间，并以此划分形成数个组团，通过围合的建筑组合营造安定宜人的邻里空间。二是南侧沿湖边布置的一排较大面积的合院住宅，是园区最高端的滨湖景观大宅。

建筑设计

建筑造型汲取传统园林中亭台楼阁的手法进行创新设计，延续古典建筑中粉墙黛瓦与自然景观之间形成的色彩搭配，并采用丰富的翼角起翘，打造灵动轻巧的建筑造型。立面取消繁复的雕梁画栋，统一提炼丰富的装饰元素，形成既符合现代审美品位，又体现中国古典精髓的建筑细节。建筑多门窗少实墙，在满足通风采光的同时，充分考虑观景、借景，增加人对室外园林的景观参与性，赏心又悦目。

Planning Design

The overall housing layout is divided into two parts: the first part is the middle courtyard groups, which are sub-divided into three zones at east, middle and west respectively. Along with meandering double cross spatial axes form a vertical and horizontal inter-connection community core space which is multilayered and image-distinctive. This zone is segmented into several clusters, and their enclosing arrangement creates tranquil and livable neighborhood spaces. The second part is the array of large house type enclosure mansions along the lake in the south, these lakeside courtyard mansions are the most upscale housing in the community.

Architectural Design

The architecture modeling adopts traditional construction techniques of pavilion, terrace and tower. As for color collocation, it inherits classical white wall and black roof-tile characteristics. Tilted eave angles make building look lithe, and facade extracts rich ornamented elements instead of complicated carved beams and painted rafters. The design not only is in line with modern aesthetic taste but incarnates Chinese classical architecture detail essence. Buildings install with more doors and windows while less walls so as to satisfy the needs of lighting and ventilation, meanwhile this is conducive to bring outdoor landscapes in and add more participatory sceneries.

景观设计

景观规划以苏州园林和街巷文化为本源，街道布局、水巷组织、庭院节点的规划都遵循步移景异的原则，营造变化流动的巷道空间和深邃悠远的景观意境，力求打造一个建筑与生态环境和谐共生的花园式社区。

规划将对北侧的城市道路进行景观改造，强化道路节点、社区围墙界面和主入口的景观设计，形成富有文化氛围和礼仪性的迎宾景观，打造高端社区的高雅形象。园区外围景观围墙高度为6.0米。在社区内部，景观设计围绕双十字形空间轴线展开，串联一系列广场、花园，造成多层次的景观变化，同时通过漏窗、月洞门、门廊和建筑错落的体型之间的互相掩映使被分隔的空间发生渗透，形成富有层次和意境的中心景观界面。

户型设计

项目采用新中式院落的空间布局，强化主庭院空间，取消消极空间，使每个院落空间都能物尽其用。对于室内空间，主卧室通过配置书房、卫生间、走入式衣柜等使其功能更完整；老人房相对独立，配起居空间和独立花园，有家庭室以供家庭活动。半室外的建筑空间加入休闲区等现代生活功能空间。

Landscape Design

Landscape planning is based on Suzhou Garden and local alley culture. The layout of streets, water lanes and courtyards complies with a principle of one step for one scene. Flowing alley spaces and vistas create a garden-like community to coexist harmoniously with ecological environments.

The planning is to reform north side street landscapes and strengthen street nodes, the surface of community wall and main entrance so as to make a ritual welcoming landscape full of a cultural atmosphere, embodying an upscale community image. The community peripheral landscape wall is 6.0 m high. In the community, landscapes creep along with double cross spatial axes, which connect an array of squares and gardens, forming multilayered landscape variation. Through ornamental perforated windows, moon gates, porticoes and dislocated buildings, separated spaces are inter-penetrated, forming layered and artistic central landscapes.

House Type Design

The project adopts a neo-Chinese courtyard space layout and highlights main courtyard space while demolishes negative space so as to make the utmost of every courtyard. As for internal spaces, the master bedroom equips with a study, a washroom and a walk-in closet to complement its functions; the old room, relatively independent, equips with living spaces and an independent garden; and there is a family room for every member enjoying family activities. Semi-outdoor architecture space adds a rest area as a modern functional space.

石库门里弄建筑 现代民国风情

杭州湖边邨酒店

Shikumen Alley Building Modern ROC Flavor

Chaptel Hotel, Hangzhou

建筑设计：北京环永汇德设计咨询有限公司 项目地址：杭州市上城区长生路57号 建造年代：20世纪30年代 建筑结构：砖木
占地面积：2 594平方米 总建筑面积：约2 268.68平方米 建筑高度：11.2米

Architecture Design: BHD Consulting Ltd. Location: No.57 Changsheng Road, Shangcheng District, Hangzhou Building Time: 1930s
Architecture Structure: Masonry-timber Site Area: 2,594 m^2 Building Area: about 2,268.68 m^2 Building Height: 11.2 m

鉴于酒店位于20世纪30年代的石库门联排里弄建筑群中，民国风情浓厚，深色木质大门的院落里，是四幢整齐的两层砖木结构的建筑。因此设计依据建筑自身特色，结合现代生活理念，以时间与空间为媒介，实现传统与时尚的共存共融。通过空间转化的方式嵌入项目独有的文化和历史元素，以德国成熟的历史建筑修缮技术还原中国传统的清水砖墙，同时采用现在的技术和设备提升居住品质，既体现了民国时期中西文化交融的时代特质，又充分展示了当代人的生活品质和时尚理念，营造出低调奢华、现代典雅的时空穿梭空间。

Chaptel Hotel situates in the 1930s Shikumen townhouses lane with thick ROC flavor. Entering a dark wooden door are four two-floor masonry-timber buildings erecting in a court. The hotel design combines its own characteristics and modern living concepts to create a place relying on time and space, uniting modernity and tradition. The design adopts unique cultural and historic elements, and takes mature German technologies to repair Chinese traditional exposed brickwork wall. Modern technologies and facilities elevate accommodation quality, which not only expresses ROC period Chinese-Western culture integration, but also presents modern habitat quality and fashionable concepts, creating an understated while luxury, modern while elegant space.

凯普江·尚品烤肉餐厅
西湖之春
假日酒店

区域分析

杭州湖边邨酒店属于全套房式小型精品酒店，位于长生路57号，步行至西湖仅需3分钟。湖边邨虽与闹市接壤，却在古砖墙的掩映之下遗世独立，散发着世纪末的洋场风情。湖边邨的特别之处，在于它以时间和空间为媒介，实现传统与时尚的共存融合。

项目背景

湖边邨建筑群坐落于长生路与蕲王路交叉口，始建于20世纪30年代，由并排的东、西两列里弄建筑组成，中间以过街楼相连。

湖边邨近代民居建筑保护区包含湖边邨、星远里、鸿源坊、大庆里、天德坊等街坊及 4 幢独立式别墅，是当时杭州规模最大、最西式的百多幢标准“弄堂房”。街区总体格局完整，建筑中西融合，巧妙自然，一式一样的两层楼，楼顶有平台、二楼有木栏杆的阳台，楼角有亭子间，其立面注重西洋建筑的雕花刻图，传统砖雕青瓦的门楣上有“湖边邨”字样，穿过马路就是西湖。这里曾居住过各式人等，是中国第一代的“白领公寓及中产阶级”的居所，亦有遗老、名人后裔等。湖边邨建筑群反映了民国时期的时代特征，作为杭州近代民居建筑的典型代表之一，继承了民族传统建筑风格，开近现代民居建筑之先风，具有极高的历史、艺术、科学、经济、社会和文化价值，是杭州城市历史发展的见证。2007年，湖边邨建筑群被市政府列入第三批历史建筑保护名单。

Location Overview

Chaptel Hotel is an all-suite small boutique hotel. Situating on No.57 Changsheng Road, it only takes three minutes to the West Lake. Although near downtown area, the hotel remains aloof from the world under the cover of ancient brick walls, emitting out foreign amorous feelings of the late 20th century. The most fascinating point of the hotel is that it takes time and space as medium to realize a coexisting of tradition and fashion.

Project Context

The Hangzhou Hubiancun building pile locates in the intersection of Changsheng Road and Qiwang Road, built in the 1930s. Two parallel lanes, in east-west orientation, compose the building pile, connected by an arcade. The Hangzhou Hubiancun modern dwelling conservation area includes Hubiancun, Xingyuan Li, Hongyuan Fang, Daqing Li, Tiande Fang and four detached villas. They were the largest and western standard alley residence at that time in Hangzhou. More than one hundred dwellings composed a complete block containing Chinese and western architecture styles. They are two-floor buildings equipping with roof terrace, balcony with wooden railings on the second floor and booth at corners. On facade, western carvings are burgeoning, and the Chinese characters “湖边邨” are carved on traditional black tile lintel. Just across a road is the West Lake. This was the first generation of white-collar apartment for middle classes in China, various people lived here including old fogy and celebrity descendants, etc. These buildings disclose ROC times features, and as one of models of Hangzhou modern dwelling, they inherit an ethnic traditional architecture style and initiate a modern dwelling trend. They have inestimable value of history, art, science, economy, society and culture, meanwhile witness Hangzhou history development. In 2007, Hangzhou Hubiancun building pile was enrolled into the third group of the Historical Building Protection List by municipal government.

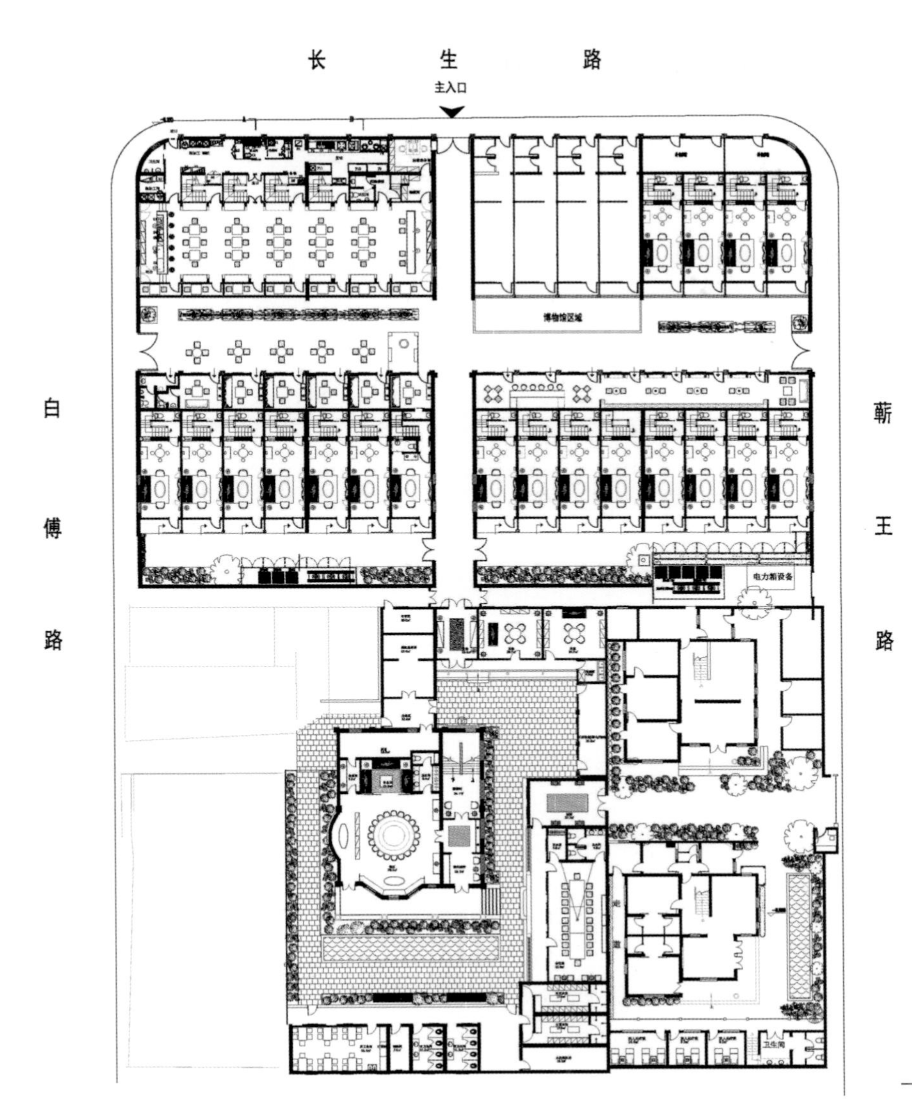

一层总平面图 1F Master Plan

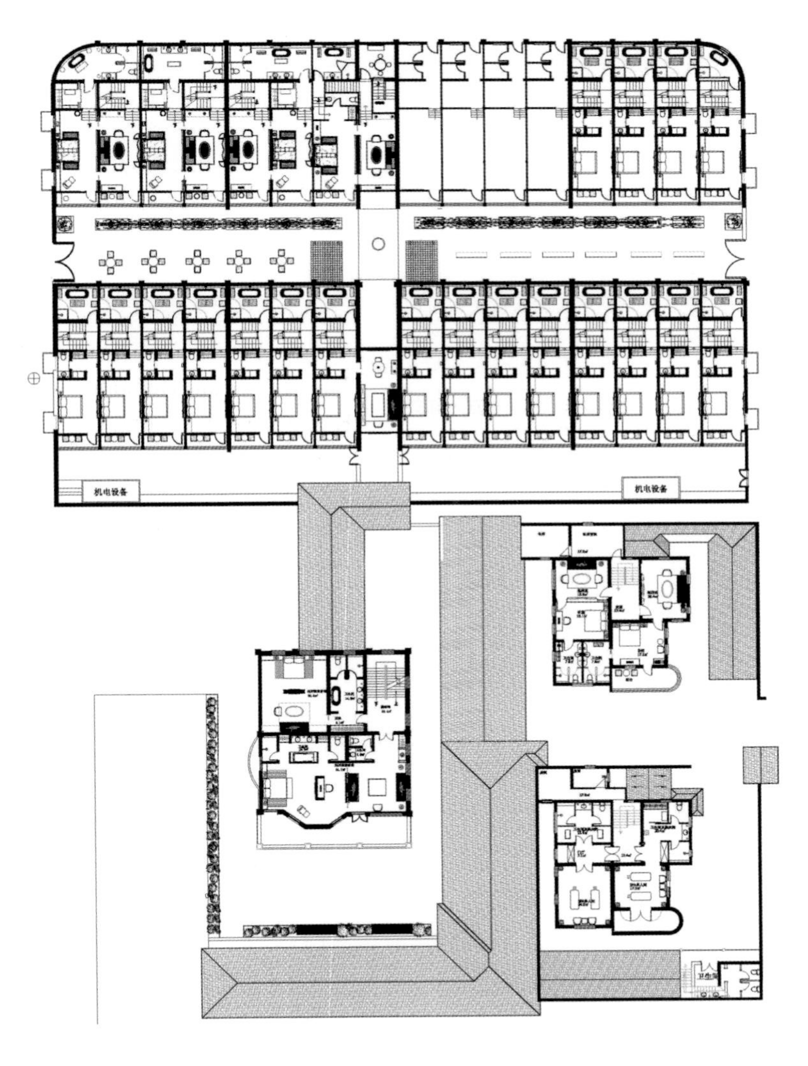

二层总平面图 2F Master Plan

项目概况

酒店主楼高2层，客房总数17间，每间配有独立阳台和晒台(即露台，石库门建筑称之为晒台)，面积均在90平方米以上。酒店内，还有一个大韩民国临时政府纪念馆，80 多年前被誉为韩国国父、开国元勋的金九先生和他的战友们组成的临时政府就在这里。该馆在原基础上扩大了3倍，分3个展厅，复原为20世纪30年代旧貌，向参观者展示当时中韩人民并肩战斗的历史篇章。

产品定位理念

酒店以城市精品酒店的概念定位、打造“城市圣所”为目标，将时间、空间融合于现代居所中，设计理念将追随新颖和历史相融合，同时满足现代化生活所需求的人性化。承诺以宾客舒适度为主旨，极力完善功能使其具备简晰与完整的设计感及生活舒适感。酒店更多基于生活方式的内涵和人文体验与它结合起来，追求至上服务与精雅设计理念的共同存在，在对民国历史时期建筑及室内风格的完整认知的基础上，以睿智的手法、成熟的解析将古典与现代、质朴与华丽、典雅与极简融为一体。酒店通过创新的理念诠释民国风情，以时尚的科技打造极致的细节，将传统与现代和谐演绎成完美，打造出一个强烈对比且又低调雅致的感观世界。

Project Overview

The Chaptel Hotel main building has two floors, including 17 guest rooms and every room equips with independent balcony and sunbathing roof (that is terrace, and the sunbathing roof is especially called in Shikumen architecture), with average area over 90 m². The Temporary Government of Korea Memorial is in the hotel. About 80 years ago, Kim Gu, the father of Korea, organized a temporary government with his companion. This memorial is enlarged three times than the former one, divided into three exhibition halls. The building restores the appearance it was, displaying the historic story that the army of Chinese and Korean fought side by side.

Positioning Conception

The Chaptel Hotel aims to build an urban holy place according to an urban boutique hotel concept. The modern place integrates time and space as well as innovation and history to satisfy modern life requirement of human-orientation. Accommodation facilities are complete and well designed as concise as possible to guarantee guests' comfortable and fashionable requirements. The hotel design bases local life style and humanistic experience, meanwhile pursues supreme service and elegance. Basing on the complete perception of ROC period architectural and interior style, the hotel design perfectly interprets classic and modern, pristine and resplendent, elegant and minimal connotations by intelligent and mature techniques. The ROC feeling is displayed by innovative concepts, and delicate details are presented by fashionable techniques. The hotel is an ultimate embodiment of tradition and modernity, because it creates a world with an intensive comparison sense, which is also understated and elegant.

小贴士

清水砖墙工艺是中西方古典建筑文化历史的重要组成部分，是独具表现力的建筑墙体装饰材料。在中国，清水砖墙的建造工艺大量应用于民居。由于青色的清水砖墙色彩饱和度很低，所以也就更容易融入自然的环境中，给人以含蓄、宁静而致远。

Tips

Technology of exposed brickwork wall is an important component of Chinese-Western classic architectural culture and history. It is a decorated material for building wall, owning unique expressiveness. In China, exposed brickwork is largely used in folk house construction. Gray is a low saturated color, so the exposed brick can easily blend in natural environment, emitting out unobtrusive and sedate feelings.

建筑设计：还原清水砖墙

青砖和木梁构建的湖边邨建筑，80多年来一直是一道古朴又靓丽的风景。由于年代久远，湖边邨原有清水墙部分砖块破损风化。为了还原记忆中的湖边邨，在给建筑做加固的过程中，尽最大努力保留湖边邨的原始风貌，设计师采用德国最成熟的历史建筑修缮方案，对湖边邨清水墙进行修缮。湖边邨酒店所属外墙每块砖石均作过修缮，同时为了建筑的整体要求又采用德国技术进行墙体的防潮处理。

Architectural Design: Revivify Exposed Brickwork Wall

Gray bricks and wooden beams construct Hubiancun buildings. During the past 80 years they have maintained pristine and pretty image. However, the exposed bricks have been gradually damaged or air-slaked as the time pass by. In order to regain the old Hubiancun in our memory, we adopt German historic architecture repairing plan and try our best to retain original appearance in the process of reinforcement. Every brick on the facade of the Chaptel Hotel is repaired, and meanwhile we employ German moisture-proof technique to protect exterior walls.

室内装饰

版画

湖边邨建筑属20世纪30年代，当时正值新兴木刻运动兴起，故此，选择版画作为酒店的装饰画，并邀请陆放老师特别为湖边邨制作了一批湖边邨版画。

家具

酒店家具均为水曲柳实木家具，按照老式家具的样式设计定制，擦色做旧，木纹明显，具有质感。酒店特有的箱几不但继承了它原有的功能，还增加了滑轮，在不用的时候将其推入桌下，腾出空间；拉出，可作凳子，翻开，可放物品。同时，酒店中也有部分家具是从老家具市场上淘来的纯正民国老家具。

灯具

灯具设计上充分吸收了Art-Deco风格，根据民国年代的锡焊手工工艺，灯具均由线条拼接而成，并加入了磨砂玻璃、背漆玻璃、热熔玻璃等元素。灯杆和开关面板保持一致，使用了紫铜做旧工艺。

Interior Decoration

Engraving Painting

Hubiancun buildings were constructed in 1930s when wood engraving activity was prevailing. That's why the Chaptel Hotel chooses engraving paintings to decorate interior spaces. In addition, painting master Lu Fang is invited to make a batch of Hubiancun engraving paintings.

Furniture

Every piece of furniture is made of ashtree solid wood, and they are in retro-pattern: antique finish with clear grain. There are special tea-trolleys can be used as seats, or hide under table to free up space, or open lid to store sundries. Some pieces are authentic ROC period furniture bought from old furniture market.

Luminaires

Luminaires adopt an Art-Deco style according to ROC soldering craftsmanship, every article of luminaire is jointed by lines, and frosted glass, painted glass and hot-melt glass are added in. Both lamp-post and switch penal are made by red copper distressed technique.

琉璃

湖边邨采用了手工琉璃的概念定制了一批大堂的艺术品，每件琉璃因为是手工定制，件件均为不同尺寸，同时专门定制了香台、烛台等器皿增加艺术点缀。琉璃的流光溢彩、变幻瑰丽的特点既体现了东方人细腻、含蓄的特点，也是湖边邨精致和品质的体现。

湖边邨的百叶

湖边邨建筑属老建筑，所以在此用了木百叶代替窗帘，符合那个年代的特征。在制作百叶过程中，考虑到传统百叶遮光性差，给百叶增加了一定的弯曲，增加遮光条，使之上下合并时能够扣紧，由此达到更好的遮光性能。

手工地毯

客厅使用的是传统手工编织剑麻地毯，卧室使用的是由80%的新西兰羊毛和20%的真丝经纯手工编织的地毯，其中手工地毯是完全采用手工编织，一针针缝制，手工地毯的特点是色泽自然，经久耐用，具有保值作用。

Colored Glaze

Artworks in lobby are handcraft colored glaze works. Due to every piece is made by handwork, they are in various sizes. Custom-made incense-stick and candle-stick add an artistic ambiance. Brilliant colored glaze features in oriental exquisite and implicit characteristics, and meanwhile presents Hubiancun delicate quality.

Shutter

Hubiancun buildings are old architecture. Shutter, instead of curtain, conforms to the decoration characteristics at that time. Traditional shutter has weak shading offects, so we remold it with a certain curve and add blind stripes to elevate shading offects.

Handmade Carpet

Living room paves handmade sisal carpet and bedroom uses 80% woolen 20% silk carpet. Every carpet is made by hand knitting, which incorporates advantages of natural color, wear-resisting and value maintenance.

苏州诗意会所 现代水墨意境

苏州姑苏会

Suzhou Poetic Club　Modern Inkwash Sphere

Soochow Club, Suzhou

规划/建筑设计：AAI国际建筑师事务所　项目地址：苏州金鸡湖西
用地面积：21 000平方米　总建筑面积：25 000平方米　采编：赵俊芳　照片来源：AAI国际建筑师事务所　摄影师：金霑

Planning/Architecture Design: Allied Architects International　Location: West of the Jinji Lake, Suzhou
Site Area: 21,000 m²　Gross Building Area: 25,000 m²　Contributing Coordinator: Zhao Junfang
Photo Source: Allied Architects International　Photographer: Jin Zhan

墙、水、径、院、桥、壁、巷、庭等元素的组合将描绘出一幅诗意的江南山水画，将苏州的小桥流水人家的意境在细节中表露无遗。姑苏会以围墙隔离喧嚣，通过厚重的大门走入宁静，通过取景与造景的方式，营造出苏州园林的体验感。同时，又利用现代材料和手法书写富有时代感的传统中式的黑白灰的水墨意境，打造出水巷邻里的现代商务会所。

Wall, water, alley, yard, bridge, screen, lane and court depict a poetic Jiangnan landscape painting. In the painting, Suzhou featured bridge, river and house are revealed entirely. Soochow Club uses wall to isolate noise and heavy chunk door to bring tranquility. Design takes view-adopting and view-building ways to create a Suzhou garden style. Meanwhile, it employs modern materials and techniques to depict a traditional Chinese black, white and gray inkwash painting while no lack of modernity. It is a modern business club in water alley neighborhood.

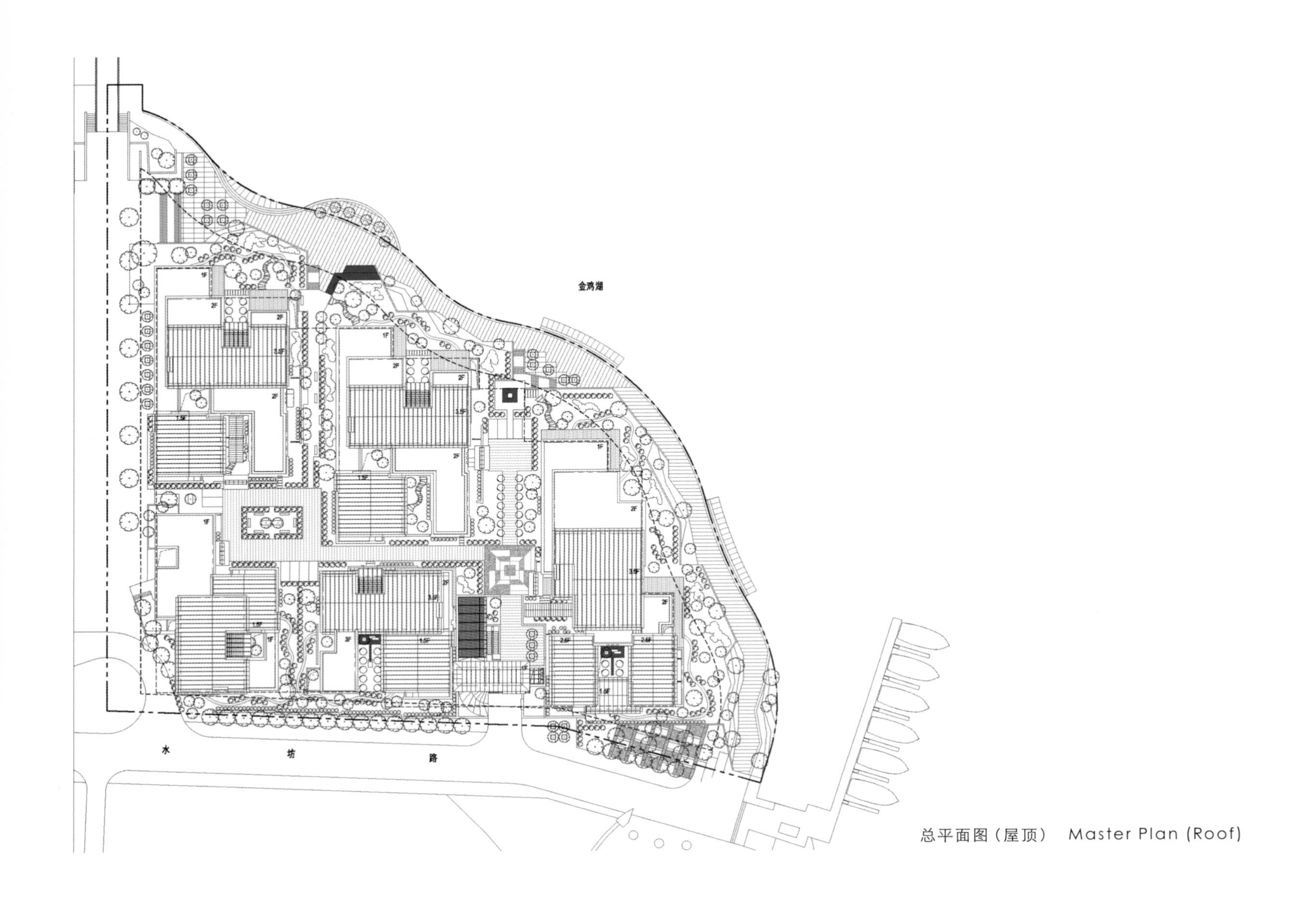

总平面图（屋顶） Master Plan (Roof)

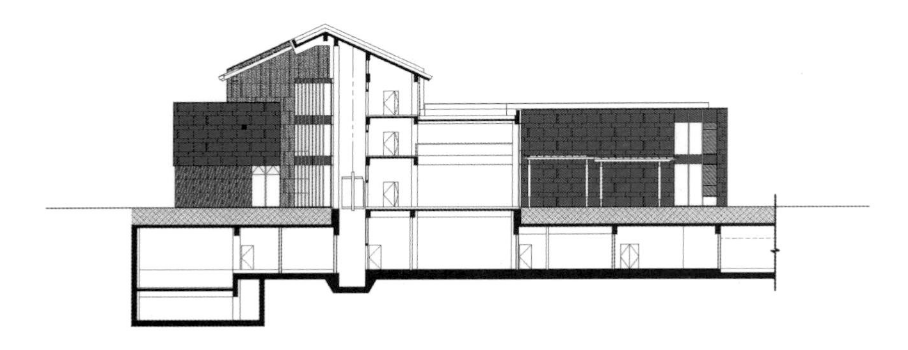

单体A 内庭院西立面图 Monomer A Inner Court West Elevation

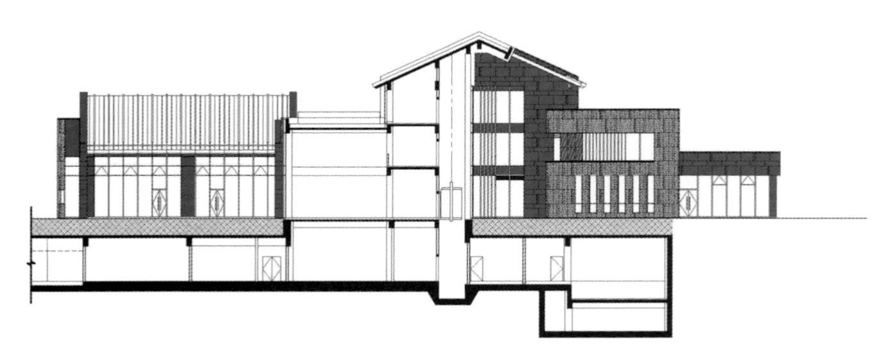

单体A 内庭院东立面图 Monomer A Inner Court East Elevation

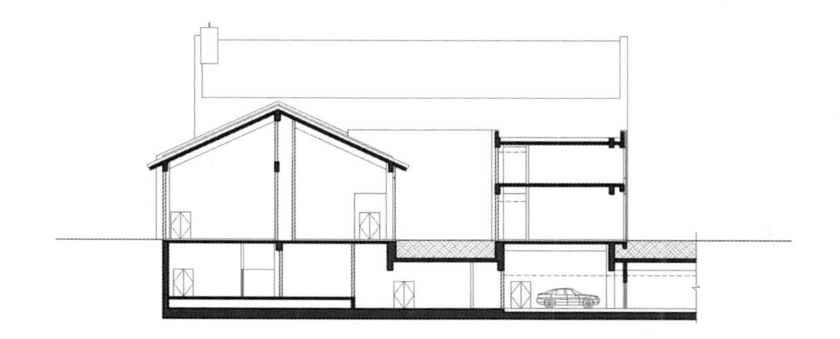

单体A 剖面图 Monomer A Section

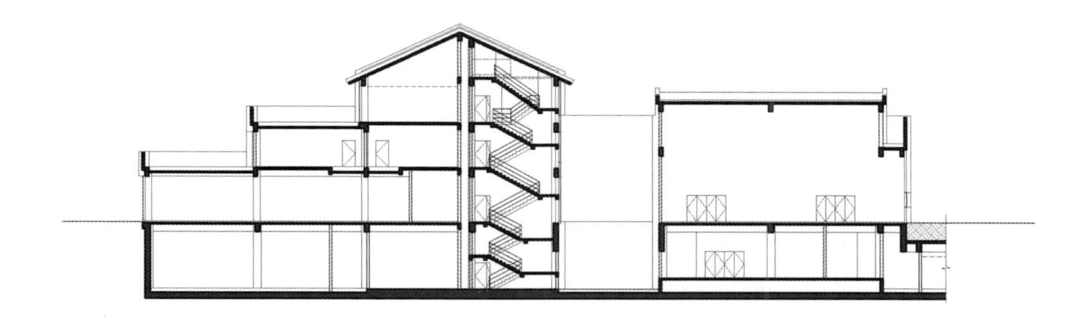

单体A 剖面图 Monomer A Section

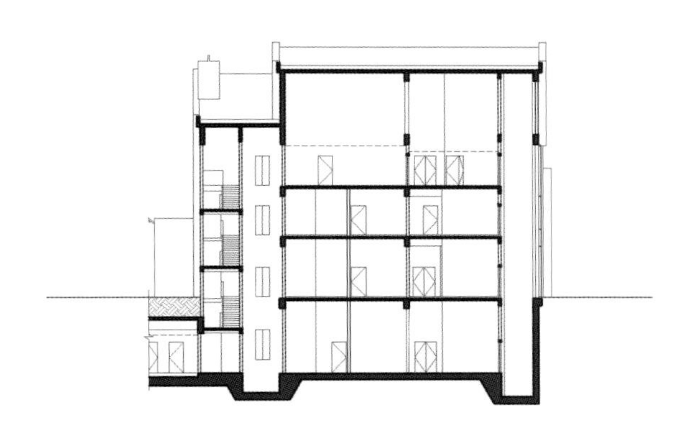

单体E 剖面图 Monomer E Section

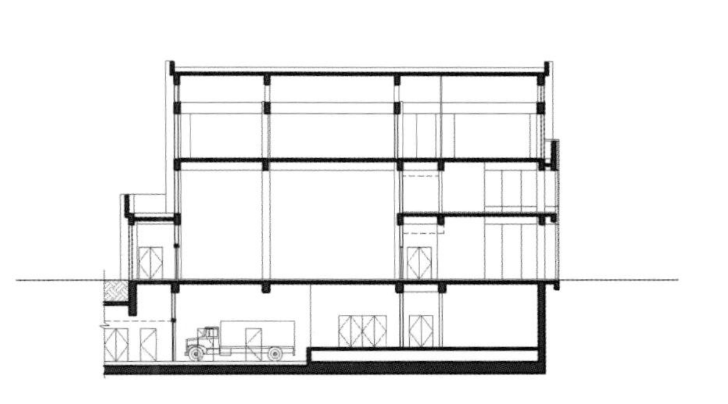

单体E 剖面图 Monomer E Section

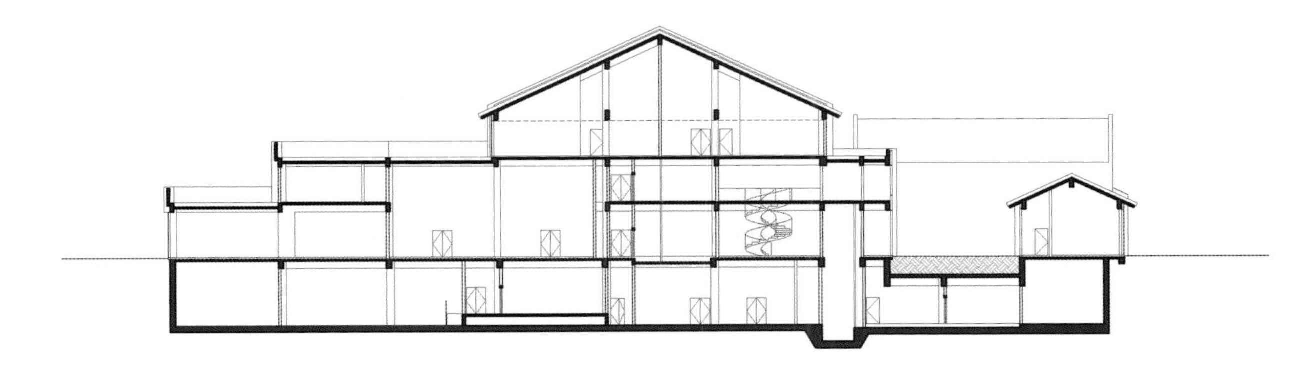

单体E 剖面图 Monomer E Section

项目概况

项目东、南两向拥有金鸡湖湖景，西侧紧邻已建水巷邻里商业区，北侧与李公堤风情水街隔湖相望，享受着得天独厚的景观资源。项目共包含5个独立的建筑，沿湖两层，余下皆为三层高，是一个顶级企业商务会所区。

Location Overview

The Jinji Lake sceneries can be seen at east and south sides of the project which is close to Jinji Lake Lane Neighborhood in the west, and Li Gong Di International Commercial Water Street across the lake in the north, enjoying exceptional landscape resources. The project has 5 freestanding buildings, and two-floor-high buildings are alongside the lake and the rest have three floors. This is an upscale commercial club.

小贴士

苏州园林又称苏州古典园林，以私家园林为主，在汉族建筑中独树一帜。而位于苏州的拙政园、留园、网师园和环秀山庄则是苏州古典园林的典范，以其意境深远、构筑精致、艺术高雅、文化内涵丰富而闻名，并充分体现出“自然美”的主旨，通过因地制宜、借景、对景、分景、隔景等手法组织空间，通过叠山水韵、花木等形成充满诗情画意的写意山水园林，对当地和周边地区的建筑文化都有着深远的影响。而在姑苏会的设计中也随处可见苏州园林的影子。

Tips

Suzhou Classical Gardens is an althernate name for Suzhou Gardens that mainly belong to private owners and develop a school of its own in Han nationality architecture. Zhuo Zheng Garden, Liu Garden, Wang Shi Garden and Huan Xiu Mountain Villa are paradigms among Suzhou classical gardens. These gardens are reputed with profound artistic conception, delicate construction, elegant art, opulent culture to express a keynote of natural beauty. They have poetic landscape feelings through adjusting measures to local conditions, view-borrowing, view-responding, view-sharing and view-blocking techniques to organize spaces and create piling rockeries and waterscapes. This kind of style deeply influences local and peripheral architectural cultures, and figures of the Suzhou gardens are witnessed everywhere in the Soochow Club.

规划设计：大隐于市

建筑整体布局上，考虑到有限的湖岸线，不仅尽力“取景”，更立足于“造景”。面积较大的三栋单体沿湖布置，利用层层退台，尽可能多地将湖景资源向内延展。基地内部较小的两栋单体，则利用建筑多变的空间组合，营造出趣味幽雅的园景，从而达到景观资源的平衡。

沿湖设置的开放式公共步道，直接连接西北角李公堤廊桥与东南角游艇码头，完成水巷邻里商业步行网络与自然公园的衔接。而场地内部与滨水步道天然存在的两米高差，足以保证基地内部景观与空间的私密性，与城市空间既相互渗透，又内外区隔。

围墙将城市的喧嚣隔离，穿过那道厚重的铜制大门走入宁静的空间，透露出大隐于市、低调奢华的人文气息。

建筑与景观：造景生情

建筑与景观亲密融合，是姑苏会建筑的一大特色。如传统苏州园林一般，建筑、景观与内部空间互相渗透、融合。除了将餐饮、会议、娱乐、住宿等功能布置其中，利用丰富多变的庭院景观提升周边功能空间的品质，建筑师更着力营造的是富有精神性内涵的建筑场所，让内部空间互相流动，给予到访者如同游走苏州园林的体验感。

通过建筑形体的变化以及对墙、院、水、壁、窗等传统元素的运用，将庭院空间划分为开放、半开放、半私密、私密四个不同级别，分别与之相对应的功能空间组合。各单体之间的景观空间也通过院墙的围合，各自收纳于建筑内部及身侧，巧妙地在“见”与“不见”之间造景生情。

Planning Design: Ensconce in City

Considering limited lakefront sceneries, the overall layout strives to “adopt views” and “build views”. Three large-size monomers are along the lake in layering retreated modeling so as to extend inward more lake scenery resources. Two small-sized monomers utilize changeable space combination to create an elegant garden so as to reach a balance of landscape resources.

Open-style public pedestrian path is along the lake, shouldering Li Gong Di at northwest corner and yacht wharf at the southeast corner. It connects up the Jinji Lake Lane Neighborhood and the Nature Park. The venue is two meters lower than the lakefront path, which guarantees a private inner space. The club creates a space not only penetrating in urban space but separating with it at the same time.

Bounding wall isolates urban noise. Through a heavy chunk copper door is a tranquil space that ensconces in city with an understated and humanistic breath.

Architecture and Landscape

The intimate relation between architecture and landscape is a great characteristic in the Soochow Club. Just like traditional Suzhou gardens, architecture, landscapes and interior spaces are mutually penetrating and merging. Besides F&B, conference, entertainment and accommodation functions, variety of courtyard landscapes elevate quality of peripheral functional spaces. Designers lay more emphasis on creating spiritual connotation and strengthening internal space communication so that visitors can have an experience of strolling in the Suzhou gardens.

Through architectural form variation and traditional elements application of wall, yard, water, screen and window, the courtyard spaces are divided into four categories: open, semi-open, semi-private and total private, corresponding with functional space combinations respectively. Landscape spaces among every monomer are enclosed by courtyard walls, folded in or drawn nearby, looming occasionally.

泰德馆

造型与材料：传统语汇现代手法

充分利用各种材料的特性展现不同的传统语汇，更是设计的精髓所在。呼应水巷邻里商业的设计理念，设计将传统元素与现代材料、手法相结合。例如深灰色亚光石灰石、白色光面珊瑚石、深灰色金属屋面及蓝灰色玻璃幕墙，既体现中式建筑黑、白、灰的水墨意境，又不失时代感。白色光面珊瑚石被刻意地划分成竖向不规则肌理，与山墙面竖向条窗结合，犹如清晨的细雨，绵绵如丝。透过观光电梯玻璃幕墙外的实木百叶欣赏庭院内风景，仿佛置身竹林深处。窗花格栅等细部装饰配以天然原木，突显“隐、雅、逸”的空间气质。

景观设计

南北向两条通透的景观主轴形成水坊路与对岸李公堤的视线穿透，将基地内部景观与金鸡湖湖景不自觉地融为一体。

同时将建筑与庭院相结合，运用设计手法在“见”与“不见”之间造情造景，化解建筑密度与景观品质的矛盾，同时与周边场地形成多个收放空间，将内部景观与城市景观相互渗透。

东侧及北侧滨水区绿地利用天然高差，结合缓坡、台阶等景观元素，使基地内部景观与城市公共景观既相互渗透，又相互区隔。

Modeling and Material:
Traditional Language and Modern Technique

The design essence of the project lies in making the best of all sorts of materials to express various traditional languages so as to echo the Jinji Lake Lane Neighborhood design concept through integrating traditional elements, advanced materials and techniques. Dark gray matte limestone, white glossy corallite, dark gray metallic roof and bluish gray glass curtain wall together reveal Chinese architectural black, white and gray flavor, flaunting a sense of modernity. Vertical irregular grain white glossy corallite echoes vertical narrow windows. Standing in sightseeing elevator, you can enjoy the landscape in courtyard through a solid wood shutter outside of the elevator's glass curtain wall, which makes you feel like standing in a grove of bamboo. Trellis windows adorned with raw wooden sculpt express a secluded, elegant and ethereal spatial atmosphere.

Landscape Design

Two transparent north-south orientation landscape main axes form a visual penetration of Shuifang Road and Li Gong Di across the lake. The project's landscapes are well integrated with waterscapes of the Jinji Lake.

Architecture and courtyards combine so well that the contradiction between architectural density and landscape quality is resolved. Meanwhile the project together with peripheral venues form free spaces to make inner landscapes and urban scenery interwoven.

Waterfront greenbelts in the east side and north side take advantage of natural height discrepancies, gentle slope and terrace landscape elements to make the inner landscapes and urban scenery interwoven and separated somewhat.

江南水乡 古镇情怀

昆山首创青旅岛尚

Jiangnan Water Town　Ancient Town Feelings

Eastern Mystery, Kunshan

开发商：首创青旅置业（昆山）有限公司　景观设计：奥雅设计集团　项目地址：昆山市锦溪镇同周公路锦商路交接处
建筑面积：416 530平方米　容积率：0.67　绿化率：31%　采编：李忍

Developer: Beijing Capital Land Kunshan Co., Ltd.　Landscape Design: L&A Design Group
Location: Intersection of Zhougong Road and Jinshang Road, Jinxi Town, Kunshan　Building Area: 416,530 m²　Plot Ratio: 0.67
Greening Rate: 31%　Contributing Coordinator: Li Ren

一堤、六坊、十三巷的规划布局，临水而筑的建筑，考究的街巷、宅门、屋顶、深墙、院落、厅堂四方围合，中式阔檐、白墙黛瓦、简洁的线条，塑造出中国传统民居的“内向型”院落空间，演绎院子的经典符号属性和传递的精神意境，传递出江南古镇的水乡情怀。并将江南诸景用现代的手法组织深化，同时连通古镇区，营造返璞归真的水乡体验，营造出烟雨江南，漠漠水乡的隐居生活氛围，将昆山的历史文化与江南水乡多元文化演绎得淋漓尽致。

One dike, six lanes and thirteen alleys layout, waterfront buildings, exquisite alleys, mansion doors, roofs, high walls, courtyards, enclosed quadrangle halls, Chinese broad eaves, white walls, black roof-tiles and concise lines together create a Chinese intra-ward courtyard space of traditional dwelling. Courtyard's classical symbols and expressed spirits deliver Jiangnan Ancient Water Town feelings. Jiangnan scenes are organized by modern techniques, and the scenes connect to ancient town areas, creating a special experience of tracing back to ancient times. The scenes also bring a misty raining secluded living atmosphere in crisscross water lanes, and fully display Kunshan history and culture, as well as a multicultural Jiangnan Water Town.

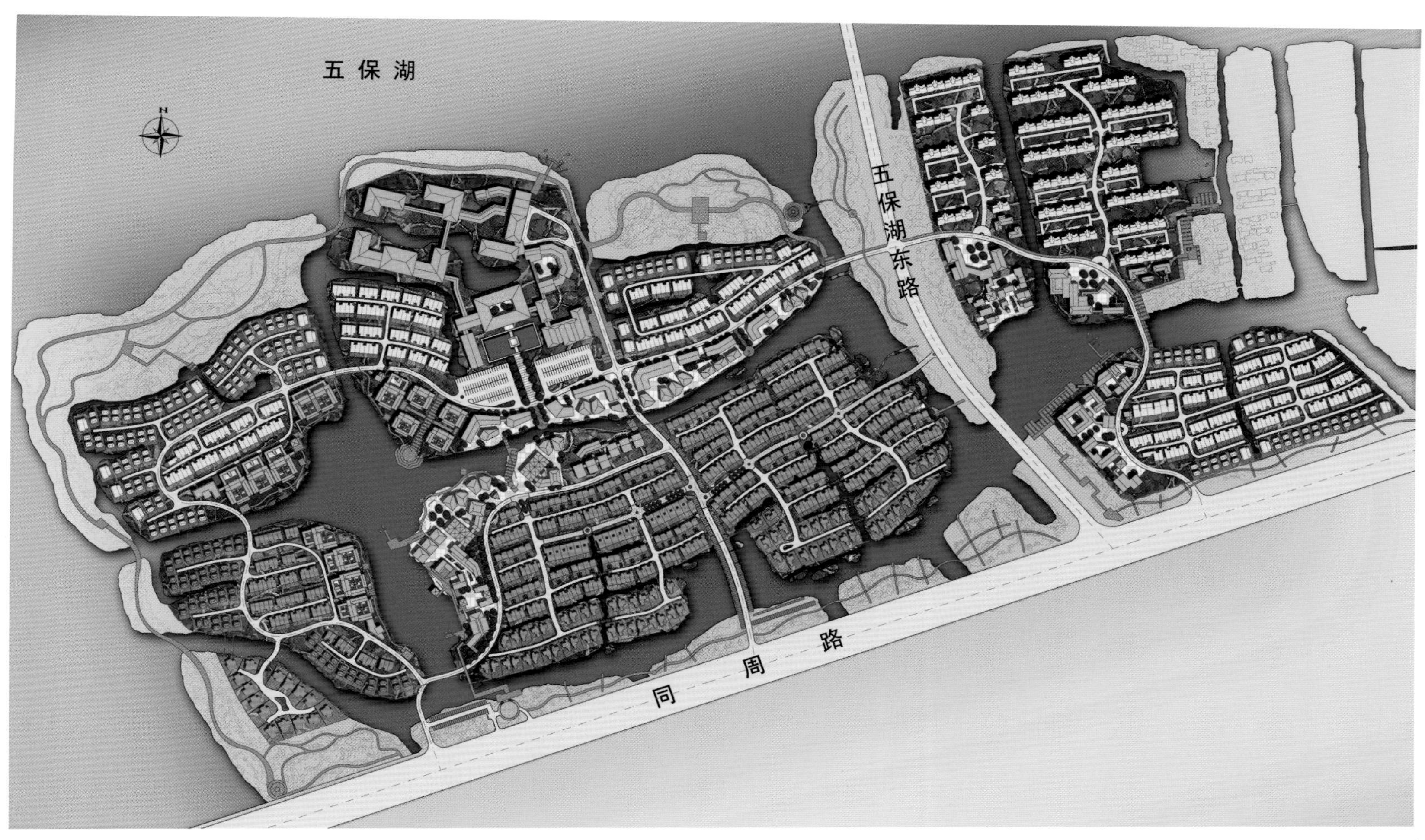
五保湖
五保湖东路
同周路

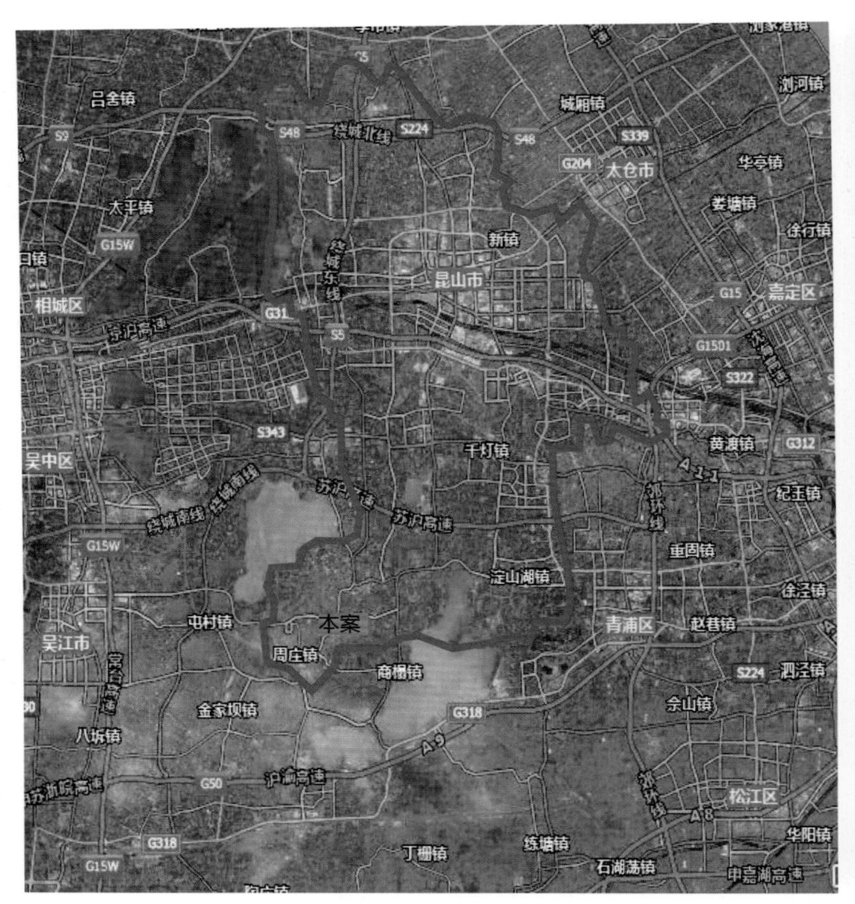

区位概况

项目位于昆山市西南区域，在锦溪镇五保湖南岸，靠近周庄古镇，与锦溪古镇隔湖相望。东临淀山湖，西依澄湖，南靠五保湖，北有矾清湖、白莲湖，拥有得天独厚自然和文化依托。

项目背景

昆山，苏州市下辖县位于江苏省东南部，水域面积占23.1%。地处上海与苏州之间；西与苏州相城区、吴中区、苏州工业园区接壤，南部水乡古镇周庄镇与吴江区毗邻，通达浙江，自秦代置娄县至今已有2 200多年的历史。

考虑到项目地处江南地区，流水景观多样，在历史长期发展过程中，形成了“小桥流水人家”的别样景致，而项目基地亦是流水环绕，流水潺潺。景观设计的愿景定义为“水乡印象”：饶镇而过的河流，水阁木楼，临河而起，绰影幢幢，小巧轻卧，杨柳依依，高高的垣墙夹着狭窄曲折的街巷，延伸向前，消失在烟雨中。

项目定位

作为古镇旅游休闲文化养生综合体开发的一期项目，项目不仅着眼于江南水乡意境与宜人岛居空间的打造，更在于对传统中式美学与文化的弘扬与传播。从而使人们放下尘埃，回归自然，找寻到梦里的山水田园。

Location Overview

The project locates at southwest of Kunshan, south shore of the Wubao Lake in Jinxi town, and close to Zhouzhuang Ancient Town, facing Jinxi Ancient Town across a lake. It looks the Dingshan Lake in the east, adjacent to the Xiyicheng Lake in the west, against the Wubao Lake in the south, and connects to the Fanqing Lake and the Bailian Lake in the north, enjoying unique natural and cultural deposits.

Project Context

Kunshan is a town subjected to Suzhou, and locates in the southeast of Jiangsu. Water area covers 23.1%, sandwiched by Shanghai and Suzhou. Kunshan connects to Xiangcheng District, Wuzhong District of Jiangsu and Suzhou Industrial Park in the west; southern water town Zhouzhuang is adjacent to Wujiang District and Zhejiang. Kunshan has more than 2,200 years since Qin dynasty when Zhilou county was built.

The project locates in Jiangnan which enjoys plenty of water resources. In the long history development, here has become a scenery of bridge, rivers and waterfront dwellings. The project is surrounded by flowing rivers, so the landscape design takes “water town image” as theme. Rivers move around the town, and terraces and wooden towers are built by the rivers with ethereal willows accompany; high walls defend long and narrow meandering lanes, stretching far away and disappearing in misty rain.

Positioning

As a project covering ancient town tourist, entertainment, culture and health maintenance, it not only concerns Jiangnan water town feelings and livable island habitat spaces, but also emphasizes promotion and propaganda of traditional Chinese aesthetics. This is an ideal place for people to unload worldly complex and return to nature.

设计愿景

项目以全新的设计思路来探索传统意境的现代演绎，从项目地域历史和江南水乡的多元文化中挖掘题材打造设计的独创性，并试图在景观概念上有所突破，着眼于水乡意境，庭院生活的营造。在环境营造上不单单满足人的观赏性，更多的是表达人对更高层次的精神诉求，超脱于尘世，使环境和心灵都达到放松与净化。

规划布局

项目规划建设478户别墅，产品类型包括双拼别墅、联排别墅，设计有4种户型，分别为建筑面积为264.95平方米的“园林式”房型；建筑面积为159.97平方米“庭院式”房型；建筑面积为169.46平方米“里弄式”房型；建筑面积为110平方米与120平方米的“叠院式”房型。园林规划“一堤、六坊、十三巷”，取意江南古镇，布局时，有机联动13条特色巷与6个景观组团，形成特色公共景观系统，体现对自然的尊重。此外，项目为住户打造全球罕见的观景平台、私家游艇码头，畅享湖居生活。岛上独创的环水设计将住区与古镇区连通，在自家门口便可体验江南水乡返璞归真的感觉。

Design Vision

The project uses brand-new design ideas to explore traditional conceptions. Designers excavate design inspiration from history and culture of Jiangnan Water Town, and meanwhile lie on water town flavor and courtyard habitation instead of common landscape design conception. As for environmental construction, it not simply pursues spectacular sceneries, but concerns higher spirit pursuit and aloof mortal life so as to release emotion and purify our soul.

Planning Layout

The planning is to build 478 villas, including deluxe villa and townhouse, with four house types and they are: 264.95 m² garden-like villa, 159.97 m² courtyard-like villa, 169.46 m² lane-and-alley-like villa and 110 m² to 120 m² folded villa. The garden takes one dike, six lanes and thirteen alleys ancient town layout. The thirteen featured alleys and six lanes are organized to form characteristic public landscapes, showing respect to nature. In addition, the project has a rare global observatory terrace and a private yacht wharf available for experiencing island habitation. The water-circular design makes the project connect to ancient towns, which offers a chance to experience Jiangnan Water Town feelings of return to nature.

小贴士

内向型院落布局是中国传统民居建筑中常见的布局方式，四面房屋各自独立，彼此之间有连廊连接，对外只有一个街门，具有很强的私密性。院落阔绰，可供居住者休闲娱乐使用，院内可种植花草树木，营造一种清雅空灵的意境。

Tips

Inward courtyard layout is a common spatial arrangement in Chinese traditional dwellings. It usually has buildings at four sides, which are connected by corridors one another, and only leaves a gate facing out, so it is conducive to form a private space. The enclosure courtyard is as large as available for leisure and entertainment, or plants flowers and trees to create an elegant and inspirational feeling.

建筑设计

项目建筑临水而筑，秉承中式建筑庭院风格，通过考究的街巷、宅门、屋顶、深墙、院落、厅堂四方围合，层层递进，实现了中国传统民居的“内向型”院落空间，通过中式阔檐、白墙黛瓦、简洁的线条，传递中式文化情怀。整体设计完美地把握了中式建筑的灵魂，将院子的经典符号属性和传递的精神意境在项目中进行全新施展。

景观设计

项目除了坐拥自然景观，还着力规划了两条主力景观轴，新中式古典园林沿轴分布，在借鉴古典园林的基础上，通过生态、场所精神、文化设计创造景观空间，重塑中国传统园林艺术中“天人合一”的智慧。

景观设计从地域出发，依托于江南水乡这个大环境，打造精致、舒适、私享的水乡小镇。追寻梦里的水乡生活，潜隐于江南水巷中，景观设计把江南诸景用现代的手法组织深化，景观元素突出“水乡的苑，水乡的街，水乡的巷，水乡的桥，水乡的码头”，营造出烟雨江南，漠漠水乡的隐居生活氛围。

植物设计围绕着中式、自然、生态等几个方面展开，注重空间营造收放相结合，注重各种植物的质感和色彩的对比，利用植物四季变化来营造自然、舒适且不一样的景观效果。此外在有限的宅间巷道空间，结合入户空间，配植中国园林中颇具典故的传统植物类型，如竹子、山茶、金桂、玉兰和海棠等，赋予空间无限的人文意境。

Architectural Design

The project builds by waters, and it inherits Chinese courtyard style. Through the progressing of exquisite alleys, mansion doors, roofs, high walls, courtyards, enclosed quadrangle hall, it realizes a Chinese traditional dwelling inward courtyard space. Chinese broad eaves, white walls and black roof-tiles express Chinese cultural feelings. The overall design grasps Chinese architectural soul through courtyard's classical symbols and spirits,which are displaying a new appearance in the project.

Landscape Design

Besides existing natural resources, two planned main landscape axes are designed in neo-Chinese classical garden style along with garden axes. Basing on classical gardens, the project creates landscape spaces and restores Chinese traditional gardens according to ecology, place spirit, and cultural design, displaying an art wisdom “unity of heaven and human”.

The landscape design relies on regional features in Jiangnan Water Town to create an exquisite, cozy and private water town. Jiangnan sceneries are organized by modern techniques, highlighting garden, alley, bridge and wharf and bringing a misty raining ambiance and a crisscross water lanes secluded living atmosphere.

Plant design focuses on Chinese style, nature and ecology themes, and pays emphasis on spatial distribution, plant texture and color comparison. Plant changing reflects four seasons transformation. In limited alley spaces and entrance spaces, Chinese traditional plants such as bamboo, camellia, thunbergii, and begony bring an unlimited humanistic reverie space.

追寻民国印记 还原历史片段
南京颐和公馆酒店
Retrospect ROC Footprint　Restore Historical Fragment
The Yihe Mansions, Nanjing

设计公司：环永汇德建筑设计咨询有限公司　主设计师：张光德
项目地址：江苏省南京市鼓楼区江苏路3号　总占地面积：约20 000平方米　采编：陈惠慧

Design Company: BHD Consulting Ltd.　Chief Designer: Zhang Deguang
Location: No.3 Jiangsu Road, Gulou District, Nanjing, Jiangsu　Site Area: about 20,000 m²
Contributing Coordinator: Chen Huihui

南京十二片区是一个相对保留完整的民国建筑群，每一栋建筑都保留其独特的历史印记，项目26栋建筑在梧桐树的掩映下仍散发着曾经的历史光晕。设计尽最大可能对原有建筑进行修缮，活化再利用原有别墅的院落形态、建筑特征，通过木质的大楼梯、百叶窗扇以及新艺术运动时期的一些建筑图腾和模纹等元素以及以现代的构造工艺还原神似的老建筑门窗上玻璃镶嵌的工艺做法，配合展览馆以及一些室内设计的细节处理，再现那个时代的印记。此外，结合一些现代设计方法和元素，形成过去与现代的穿梭，营造出具有浓郁民国氛围的现代艺术酒店，增强体验感。

The project locates in Nanjing Twelve District where conserves relatively intact Republic of China (ROC) period buildings, and every building in the district owns individual historical memories. The project includes 26 monomers distributing historic halo under shading of sycamore trees. The design attempts to maintain and repair the old buildings, and recovers original villa courtyard formation and some characteristic architectural components: grand wooden stair, louver casement, architectural totems and patterns in Art Nouveau period. Modern techniques make replica of glazing on old windows and doors, and museum displaying and interior details together make us return to ROC times. The combination of modern and traditional methods and elements brings a time travel space, creating a ROC style modern art hotel.

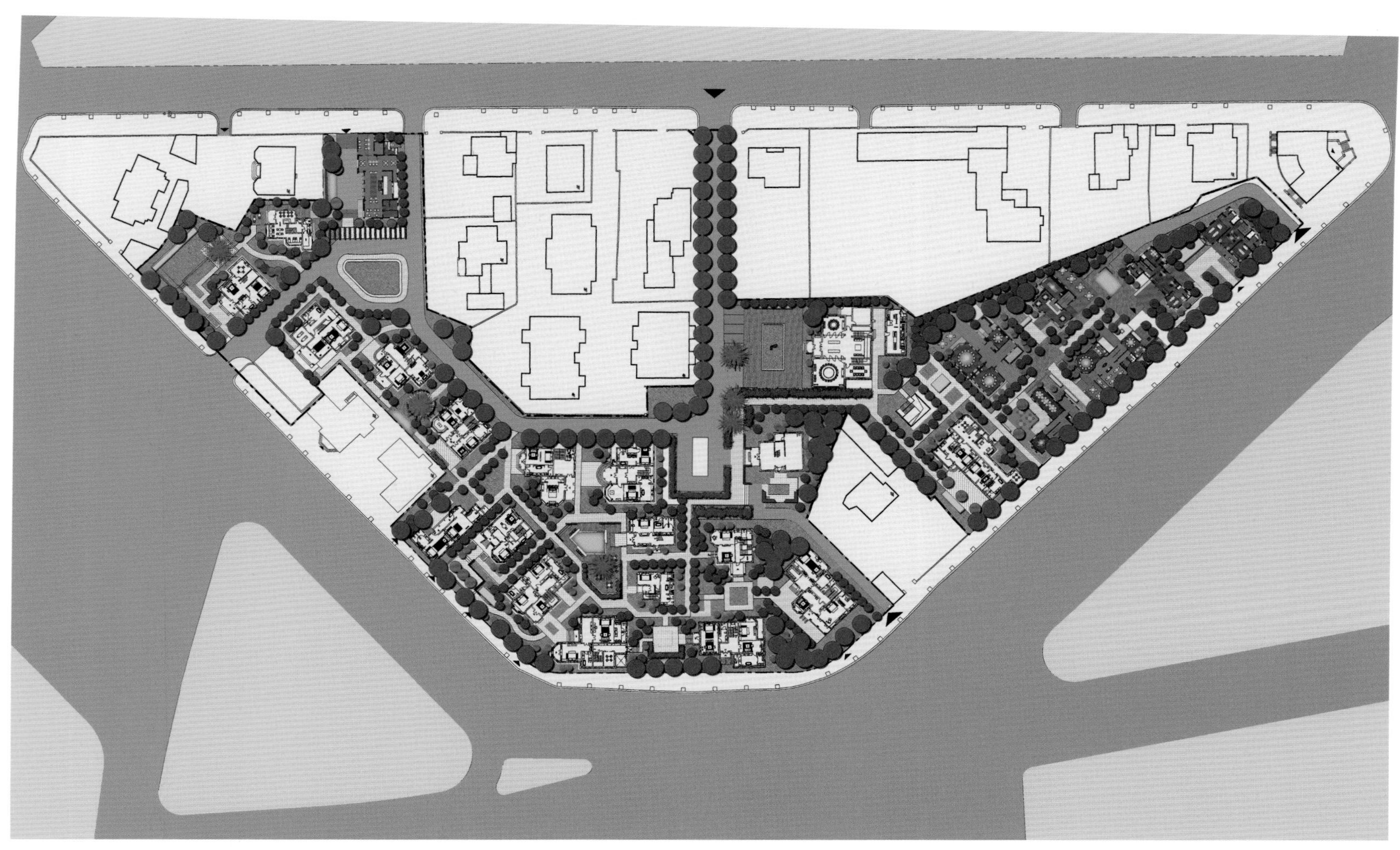

区域分析

南京颐和公馆精品酒店位于南京市鼓楼CBD的颐和路公馆区，即颐和路与宁海路交界处，毗邻江苏省委省政府所在地。现存26幢独立的民国时期的别墅建筑，属于民国原首都规划的一部分，是南京重要的历史文化街区。

项目概况

酒店建筑为现存26幢独立的民国时期别墅建筑，其中涵盖大床间、套间、独栋等30余间客房；餐厅包含中西餐，契合民国首都规划建设为中西合璧的历史背景，也满足了客人的需求；酒店还设置了多功能会议中心，满足精品酒店的基本诉求。此外，还设有文化博物馆、民国风格的舞厅，使客人可以身临其境地感受现代南京及其相关的民国历史。

Location Overview

The Yihe Mansions locates in Nanjing Drum-Tower CBD, Yihe Road Mansions zone, the intersection of Yihe Road and Ninghai Road where is close to the Jiangsu Provincial Government. This area has 26 freestanding ROC period villa buildings which belonged to a part of planned former ROC capital, and now it is an important historic and cultural block in Nanjing.

Project Overview

The Yihe Mansions consists of 26 independent ROC period villa buildings, amount to over 30 guest rooms including double room, suite and detached room. Dining hall provides Chinese meal and Western meal, which echoes historical background of ROC capital planned architecture featuring in combination of Chinese and Western, and meanwhile meets customer's requirements. The hotel equips multifuctional conference rooms and SPA facilities to satisfy basic demands of boutique hotel. Moreover, it has cultural museum and ROC style ballroom for customers to experience modern Nanjing and Nanjing ROC history.

小贴士

南京颐和公馆所在的十二片区作为南京著名的“使馆区”，是一个相对保留完整的民国建筑群，留存有相对较好的建筑风貌。这里曾经作为民国时期一代叱咤风云人物的官邸，每一栋建筑都有他独特的历史印记，缩影出当时最经典的建筑形态，同时，也成为南京这座城市历史记忆的留存。

Tips

Nanjing Yihe Mansions settles in “Embassy District”, the Nanjing Twelve District conserving relatively intact ROC architecture buildings. Many mansions here belonged to celebrities in ROC period, so every mansion owns individual historical memories. These mansions are epitomes of classic architectural type at that time, and become precious historical relics in Nanjing.

设计定位

设计定位以保护性修缮为主，提取全新的居住体验形式移植到老建筑群落当中，使客人在这里除了拥有舒适的居住体验，还可以完整地体验历史、体验城市、玩味记忆。从而既可以保护城市历史文脉，又可以激活发展城市老区，从而打造南京新的城市名片。

规划设计

项目设计通过研究26栋单体建筑之间的关系，尽可能地把这26栋风格和色彩均不相同的建筑有机地整合到一起，呈现出景观围墙、水体、绿化、照明以及新增的门楼全部和这些原有建筑群落融为完美一体的建筑形态，并在这些零散的元素中寻找共性，加以提炼，贯穿整个园区。

建筑设计

在设计中，设计师通过搜寻大量的资料对设计进行梳理和提留，希望尽可能地保留那个时代的一些建筑印记，如木质的大楼梯、百叶窗扇以及新艺术运动时期的一些建筑图腾和模纹等。再如通过现在的构造工艺做出模仿神似的老建筑门窗上玻璃镶嵌的工艺做法。对于比较有名的薛岳和陈布雷故居，从颜色和形式符号上去渲染他们一文一武的主人特点。

Positioning

The design mainly focuses on maintenance and repairing. In old building groups, brand new living experience is brought in to offer customers with comfortable residing feelings. Customers also can experience local history and urban spaces, and retrospect historic footprints. The project not only protects urban historical context, but activates development for the old town area. The mansions district will become a new name card in Nanjing.

Planning Design

Relation of the 26 monomers is carefully studied to organically combine the different styles and colors. Landscape wall, waterscape, greening, lighting and gatehouse are utilized to link old buildings into a unity. Designers try to find common characters among scattered elements, and then refine and distribute them in the whole building groups.

Architectural Design

Designers searched and reviewed many historical data to restore architectural style at that time, such as grand wooden stair, louver casement, architectural totems and patterns in Art Nouveau period. Modern techniques make replica of glazing on old windows and doors. Such as historical celebrity Xue Yue and Chen Bulei' former residences, color and symbol are used to render their characters of literatus and valiant respectively.

室内设计

酒店大堂入口设于原公馆区的内部街道，来到酒店，便置身于民国建筑群间，客人可以感受到鲜明的历史文化气息，想象着薛岳将军、黄仁霖、陈布雷等历史名人的事迹。酒店客房设计嵌入项目唯有的历史与文化元素，特别是现代南京及其相关的民国历史，部分客房参考建筑原主人生平进行设计，保证文化内涵的同时满足多样灵活的入住需求。

豪华双人间客房面积约48平方米，通过木质楼梯、拼花地板、老虎窗塑造浓郁的民国气息，与智能化房间感应科技完美融合，配以高档的纯天然埃及长绒棉床品及美国丝涟品牌床垫，营造温馨舒适的入住体验。

豪华单人间面积约为42~60平方米，配以水曲柳实木家私及高档白玉釉卫浴瓷，进口美国丝涟品牌床垫和纯天然埃及长绒棉床品，轻松打造奢华精致的隐逸之所。

公馆豪华套房藏身于多座民国官邸之中，拥有60~80平方米的舒适空间，套房内全部采用简约复古的水曲柳实木家私，打造超凡的入住体验。

公馆特色套房面积约为45~70平方米，设计错落优雅，营造不一样的温馨氛围。

Interior Design

The hotel entrance hall sets at the internal street of the mansions district. Customers will be surrounded by ROC period architectural buildings when they arrive at the hotel. They can feel a distinct historic and cultural ambiance. Guest room design introduces elements of modern Nanjing culture and Nanjing ROC history. Some guest rooms adopt the former owner's layout. The design aims to hold cultural connotation, while meets various accommodation requirements.

Deluxe Double Room is about 48 m^2. Wooden stair, parquet and dormer reflect ROC featured style. The room perfectly integrates in intelligent interaction techniques, high-quality natural Egypt long-staple cotton bedding and American Sealy mattress, creating a comfortable accommodation experience.

Deluxe Single Room is about 42 m^2 to 60 m^2. Ashtree solid wood furniture, upscale white jade glaze sanitary porcelain, imported American Sealy mattress and natural Egypt long-staple cotton bedding make a delicate while recluse space.

Mansion Deluxe Suites hide in many ROC mansions with 60 m^2 to 80 m^2 area. Every piece of furniture is made of ashtree solid wood. Customers will have a cozy accommodation experience.

Mansion Characteristic Suite is about 45 m^2 to 70 m^2 with elegant decoration, bringing a unique warm and sweet atmosphere.

古今融合 扬州特色街区

扬州虹桥坊

Ancient and Modern Syncretism Yangzhou Charateristic Block

Rainbow Square, Yangzhou

开发商：扬州瘦西湖旅游发展集团有限公司　建筑设计：上海都设建筑设计有限公司　配合单位：江苏时代建筑设计院有限公司　景观设计：上海都设建筑设计有限公司
配合单位：扬州园林设计院有限公司　项目地址：江苏省扬州市广陵区大虹桥路18号　占地面积：48 000平方米　地上建筑面积：25 000平方米
容积率：0.52　绿化率：32%　摄影：凌克戈　采编：赵俊芳

Developer: Slender West Lake Tourism & Development Group　Architecture Design: Dushe Architectural Design
Architecture Cooperator: Jiangsu Provincial Architectural Design & Research Institute　Landscape Design: Dushe Architectural Design
Landscape Cooperator: Yang Zhou Yuan Lin She Ji Yuan　Location: No.18 Dahongqiao Road, Guangling District, Yangzhou, Jiangsu
Site Area: 48,000 m²　Building Area above Ground: 25,000 m²　Plot Ratio: 0.52　Greening Rate: 32%
Photographer: Ling Kege　Contributing Coordinator: Zhao Junfang

瘦西湖有着深厚的历史文化底蕴，被列为世界文化遗产，其历史文化性投映在周边建筑中，形成了独特的建筑风情。扬州虹桥坊设计从区域的历史文化着手，定位为瘦西湖文化休闲特色街区，充分发挥瘦西湖的景观资源，采用园林式布局组成一个多动线漫游式商业街，并通过对建筑的分层处理，在二层设计中采用传统设计元素——木花窗和木板饰面，与历史风貌区相呼应。同时，在首层运用现代材料——石材幕墙构建现代商业载体，既能呈现出原汁原味的传统风貌，又透露出现代商业理念，打造出古今结合、适应现代城市休闲商业的商业街区。

Slender West Lake, listed as world cultural heritage, possesses profound historic and cultural deposits whose style radiates peripheral buildings, forming a unique architectural flavor. The Rainbow Square is just one of them, and it bases on history and culture, fixes on Slender West Lake Cultural Leisure Block, takes advantage of landscape resources and uses garden-like layout to compose a multifunctional meandering commercial street. Through a hierarchical processing method, traditional design elements are applied on the second floor-wooden lattice windows and wooden veneer to echo the history scenic area, meanwhile the first floor adopts modern material-stone curtain wall to build modern commercial image, expressing an original traditional appearance and revealing a modern commercial concept, and the rendering is a commercial block in syncretism of ancient and modern and adapting contemporary urban leisure.

Häagen-Dazs
Häagen-Dazs
Häagen-Dazs
雅珀朗丽

瘦
西
湖
11
7
8
9
10
1
2
4
6
3
5
旅游信息中心
入口广场
大虹桥
大 虹 桥 路

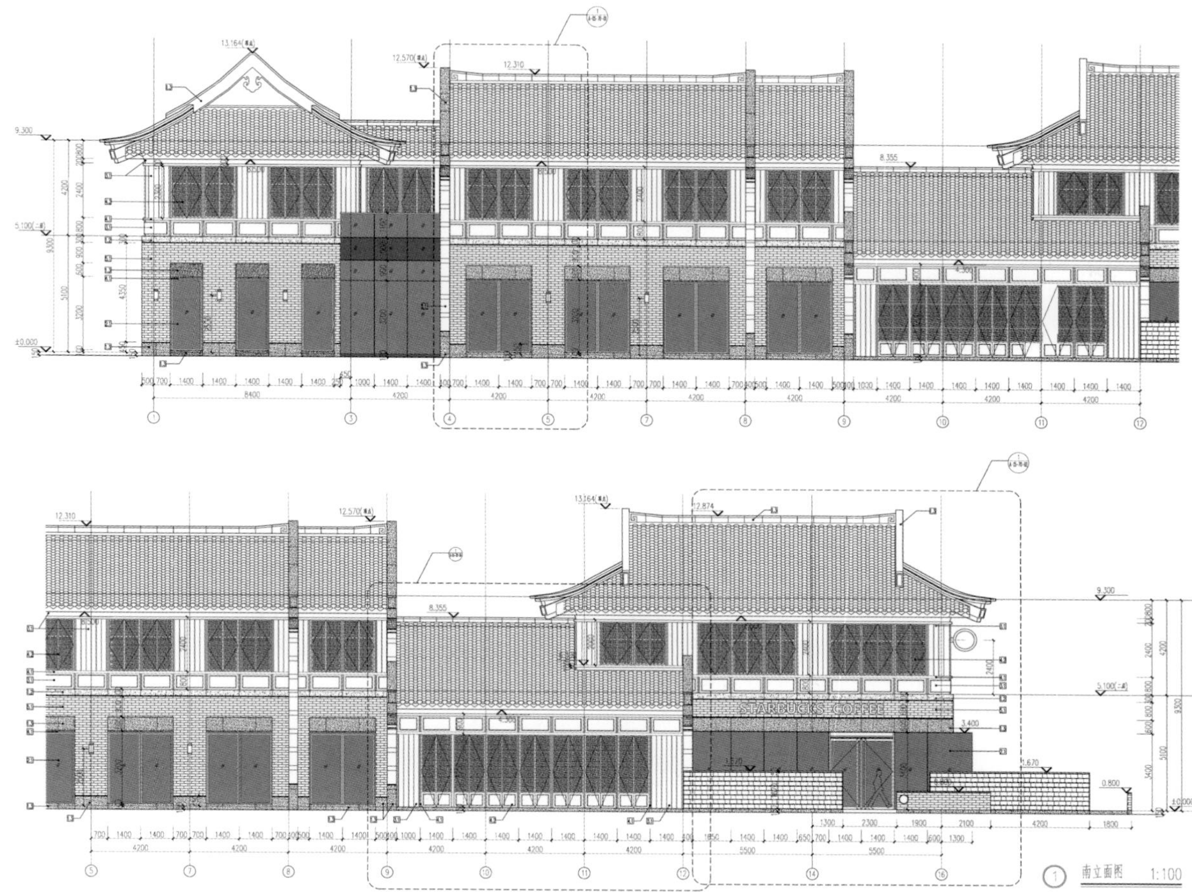

5号楼南立面 Building 5 South Elevation

区位分析

扬州虹桥坊位于扬州大虹桥路北侧，西邻瘦西湖，是瘦西湖风景名胜区的南大门。项目地处扬州核心强势商圈，坐拥绝版人文资源，交通便利。

项目背景

扬州瘦西湖，2014年被列为世界文化遗产，是我国第46个世界遗产京杭大运河的重要组成部分，是具有重要历史文化遗产和扬州园林特色的国家重点名胜区，是扬州园林的代表作。曾有诗词如此形容，“两岸花柳全依水，一路楼台直到园”，清晰地描绘出了扬州园林的择地构园的特点及所营造出的一片自然风光。

项目定位

项目整体定位为瘦西湖文化休闲特色街区，是以历史人文为主题，集餐饮、休闲娱乐、零售、展示演出为一体的一站式、全天候、现代时尚休闲商业街区。

规划设计

扬州虹桥坊为10栋单体围合而成的建筑群落，两层为主，局部三层。项目规划停车位共509个。在总体上，10栋单体按照园林式布局组成了一个多动线漫游式商业街。其中1号楼、7号楼、11号楼坐拥瘦西湖第一线湖景绝版资源。5号、6号、9号楼面向城市商业广场，将以国际品牌休闲餐饮为主，引进星巴克和哈根达斯的旗舰店作为休闲新地标。其他楼宇以城市休闲生活服务为主要业态，共同打造一个充满活力的城市商业街区。

Location Overview

Rainbow Square situates at the north of Dahongqiao Road, Yangzhou, close to Slender West Lake. It is the south entrance to Slender West Lake scenic spot. The project locates in the core business area of Yangzhou and occupies unique humanistic resources with convenient transportation conditions.

Project Context

The Slender West Lake was listed as world cultural heritage in 2014, and now it is an important component of the Great Canal, the 46th world heritage in China. The Slender West Lake bears significant historic and cultural heritage and is designated as one of National Key Scenic and Historic Interest Spots. It is a representative of Yangzhou Gardens, and there is a poem says “flowers and willows planting on both sides of the river, a meandering passage through pavilions and terraces to gardens”, which clearly points out the Yangzhou Gardens feature that garden designs with landforms and depicts natural sceneries.

Positioning

The project positions as a Slender West Lake cultural leisure block. It takes history and humanity as themes, including F&B, entertainment, retails, exhibition and fashion show. It is a one-stop and all-day-long modern commercial block.

Planning Design

The Rainbow Square is an enclosed building community with 10 monomers, and most of them are two-floors and some are three-floors, offering 509 parking spaces. The 10 monomers compose a multicirculation business street, in a garden-like layout. The building 1, building 7 and building 11 enjoy first line Slender West Lake waterscapes, and building 5, building 6 and building 9 face a commercial plaza, entered many international leisure and F&B brands, and among them Starbucks Coffee and Haagen-Dazs flagship stores become new landmarks. The rest buildings are for urban leisure and living services. This is a dynamic urban commercial block.

建筑设计

立面处理

扬州虹桥坊在设计上采用了一种折中的策略，以尽量顾全各个群体对项目的期望。设计将建筑的首层和二层分开处理：屋顶完全使用传统做法进行复原，二层以实木花窗和木板饰面为主，取传统风貌之精华；首层则采用石材幕墙做法，以大面积的展示橱窗应对现代商业的需求。这样，无论是从瘦西湖风景区内部还是从城市主干道上来看，由于树木的遮挡，给人留下印象的主要是坡屋顶和二层立面的局部，能够完全符合传统风貌的需求。而当游客通过入口广场，进入到商业街环境之中的时候，就会发现，首层干净挺括的立面正是丰富多彩的商业活动最佳的背景。首层基座和上部体量通过相同的材料、立面的划分统一起来，形成了各有特色又和谐统一的整体风貌。这种策略最大程度地综合了各种有利因素，既能符合传统风貌的要求，又具有现代商业的使用便利，为业主创造价值，也为城市创造了价值。

设计手法

项目设计坚持从传统扬州园林中汲取精华，运用现代设计语言进行提炼的设计原则，以空间和意境打动人，以细节和构造提升项目的品质。针对虹桥坊项目专门进行了技术研发。例如普通青砖墙面采用湿贴方法，天长日久容易泛碱。为了克服这一缺陷，在项目中新创了青砖干挂施工法，通过简易的干挂措施使表面青砖和基层墙面分离，杜绝了泛碱现象的出现，大大提升了项目的品质，而造价只增加了10%。建筑山墙采用石材幕墙立面，米白色木化石采用开缝构造，增加了立面细节，使得整体建筑立面十分耐看。8号楼外立面采用铝合金镂空花纹装饰幕墙配合U形玻璃，现代的形象与周边建筑形成对比，相得益彰。面向瘦西湖景区的1号楼、7号楼、11号楼，是作为高档餐饮和会所使用，建筑首层以外有传统风格的漏窗围墙进行围合，保证了首层的私密性。

Architectural Design

Facade

The design adopts a compromise strategy to cater to various expectations from different objective groups. The first and second floors have distinct designs: the roofs restore in traditional methods; the second floors use solid wooden grille windows and wooden veneers to display traditional adornment essence; the first floors take stone curtain walls to satisfy modern commercial needs for large-size display window. Due to trees shielding, seeing from the inside of the Slender West Lake scenic spot or from arterial roads, the slope roofs and parts of the second facade leaves an image that the Rainbow Square completely dwell in tradition. While customers entering into the commercial street from entrance plaza, they will find clean facades on the first floor act as the best backdrop for opulent commercial activities. The foundation bed and the upper body employ the same materials and uniform facade division, which makes the whole facade disparately while in harmonious unity. The design composites all aspects of favorable factors to the extreme so as to meet traditional style needs and modern commercial convenience, creating value for clients and the whole city.

Design Technique

The project abstracts design essence from the Yangzhou Gardens, uses modern design language to refine design principles, utilizes space and art conception to elevate affection and emphasizes details to promote quality. Designers did a special design technique research for the Rainbow Square. Ordinary black brick usually takes wet stick technique, while this way easily incurs saltpetering. In order to solve this flaw, black brick dry-hang construction technique is invented to depart the black brick surface and base wall face so as to prevent saltpetering phenomenon. It elevates the project's quality while just raise 10% cost. The gable wall chooses stone curtain, and beige woodstone with seams strengthens facade details, making the whole architecture be attractive. Building 8 facade uses aluminium alloy curtain wall with pierced pattern ornaments to match U-shape glass, and its modern image set off peripheral buildings very well. The building 1, 7 and 11 facing the Slender West Lake, are used for upscale F&B and clubs, and a circular of traditional brattice protects the privacy for the first floors.

Häagen-Dazs

满记
甜品

满记甜品
HONEY MOON DES SERT

Yechun Teahouse

STARBUCKS COFFEE

沉静怀旧民国风 秀气典雅江南景

南京宏图上水庭院

Pacific and Reminiscent ROC Style　Delicate and Elegant Jiangnan Scenery

Gentler River Villa, Nanjing

开发商：宏图地产　建筑设计：上海柏涛建筑设计咨询有限公司　景观设计：上海柏涛建筑设计咨询有限公司（仅启动区）　项目地址：江苏省南京市雨花台区软件大道
占地面积：89 600平方米　建筑面积：109 600平方米　绿化率：30%　容积率：1.22　摄影：张全　采编：谭杰

Developer: Hirealty　Architecture Design: Shanghai Botao Landscape Art & Design Co., Ltd.
Landscape Design: Shanghai Botao Landscape Art & Design Co., Ltd. (only initiating zone)　Location: Ruanjian Avenue, Yuhuatai District, Nanjing, Jiangsu
Site Area: 89,600 m²　Building Area: 109,600 m²　Greening Rate: 30%　Plot Ratio: 1.22
Photographer: Zhang Quan　Contributor: Hirealty　Contributing Coordinator: Tan Jie

南京是个历史印记厚重的城市，作为民国政府所在地，中西交融的民国建筑在南京留下了浓重的一笔。宏图上水庭院的整体建筑采用了沉稳庄重的灰砖墙体，点缀红砖、红褐拱窗，沿承金陵历史文脉，加以现代元素，形成了一种中西文化的交融，再现浓郁民国风情。并以贯穿小区南北的中心主要水系，体现江南风景的秀气。景观设计则汲取民国建筑的传统精髓，并糅合现代舒适生活居住要求，首创“背山面水”规划方式打造宁静典雅的生活空间。

At the location of old Government of the Republic of China (ROC), Nanjing remains many historic relics, especially the ROC architecture brings deep influences. The Gentler River Villa inherits Nanjing historic texture and adopts grave gray walls, red bricks and rose beige arched windows embellishing in, meanwhile adds modern elements to form an architectural combination of Chinese and Western, reappearing ROC style. A central water system transits south to north of the community, expressing a Jiangnan landscape spirit. Landscape design abstracts ROC architectural traditional essence while meets modern comfortable living requirements, initiating “facing waters and against mountains” planning and creating a tranquil and elegant environment.

二期叠加别墅
Overlay Villa

二期联排别墅
Town house

二期三进四院
Courtyard

三期小高层
High Rise

商业
Commercial

上水总平面图 Master Plan

二期规划布局图 Site Plan

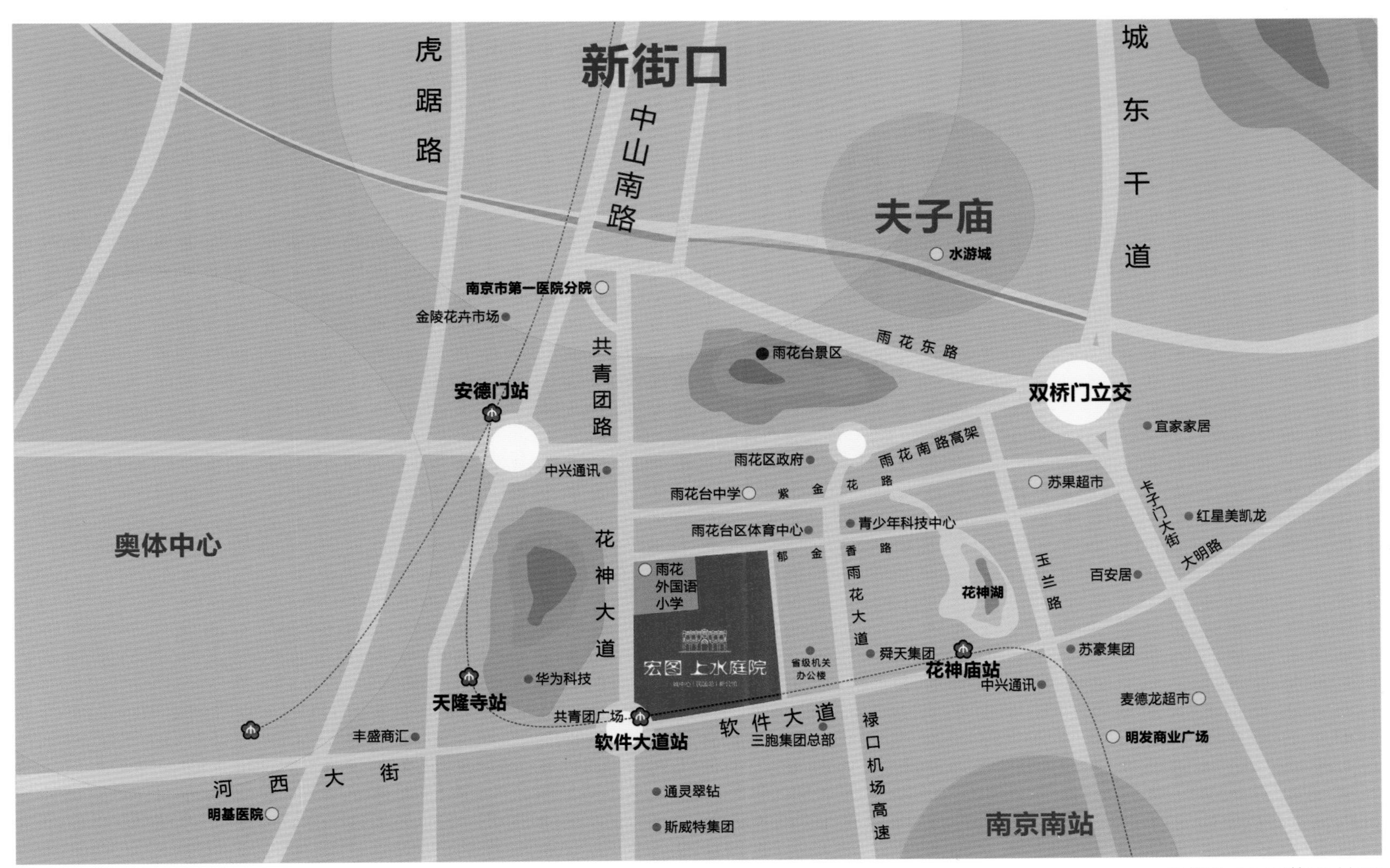

区位图 Location Map

区位分析

项目位于城南板块核心位置，南面为软件大道，西面为花神大道， 北面为郁金香路，地铁一号线（软件大道站点）毗邻小区门口，两站即抵南京南站，6站直达新街口，机场高速、绕城公路等多条快速通道环伺。项目周边配套成熟、高科技企业林立，区域内居住的皆是大型企业高素质人群。

项目概况

整个项目占地面积约15万平方米，总规划建筑面积约28万平方米，总建筑面积约8万平方米，主要由叠加别墅（16幢140套）、联排别墅（8幢70套）、三进四院（1幢9套）组成，并且配建有幼儿园、社区中心、商业街区，为您的生活提供全面的生活配套。

Location Overview

The project situates in the core area of south in Nanjing, with Ruanjian Avenue in the south, Huashen Avenue in the west and Yujinxiang Road in the north. Subway line 1 Ruanjian Avenue Station is adjacent to the community entrance, just two stations from Nanjing South Station and six stations from Xinjiekou Station, and the Airport Expressway and other speed ways are around the community as well. The peripheral supporting facilities are complete, and high-tech companies are standing by. It is an upscale villa community, accordingly most of the residents are from large-scale enterprises with high quality.

Project Overview

The whole project's site area is about 150,000 m², planned gross building area 280,000 m², and gross floor area 80,000 m². There are 16 overlay villas together 140 suites, 8 townhouses together 70 suites and 1 quadrangle courtyard equipping three entrances together 9 suites. The community offers complete living supporting facilities, such as kindergarten, community center and commercial street.

小贴士

南京是个历史痕迹厚重的城市，南京的民国建筑是中国近代建筑（1840~1949）的一个重要组成部分，经历了由照搬照抄到洋为中用的发展轨迹。时至今日，南京的大部分民国建筑保存完好，构成了南京有别于其它城市的独特风貌。南京民国建筑，其设计、构造、风格，既体现了近代以来西方建筑风格对中国的影响，又保持了中国民族传统的建筑特色，全面展现了中国传统建筑向现代建筑的演变，它在建筑发展史上具有重要的意义。文物代表着一个城市深刻的文化内涵，显示一个城市的文明程度。山水城林、历史文物和现代建筑，交相辉映，正是南京的重要标志。

Tips

Nanjing remains many historic relics, and the ROC period buildings are a crucial part of modern history (1840–1949) architecture. Architectural design has experienced two development periods: copying western and adopting selectively. Most of ROC buildings have been well conserved in Nanjing, which is conducive for Nanjing to form its unique appeal differ from other cities. From design, configuration and style of Nanjing ROC architecture, we can know modern western architectural style and how to influence China and maintain Chinese ethnic traditional architectural features. These relics unfold a scroll of painting that show a process of Chinese traditional architecture developing to contemporary architecture. The remaining buildings possess profound historic significance in architectural development. Historical relics indicate a cultural connotation and civilization level of a city. Urban landscapes, historical relics and modern buildings interplay, forming Nanjing significant symbols.

设计主题：沉静怀旧的民国风

项目力图丢弃拷贝式的追随潮流，挖掘地域性产品的独特属性，提取南京传统民国建筑的造型元素，塑造项目可识别的整体风格。通过灰砖的主墙面、少量红砖的点缀、灰瓦屋面、深灰褐色窗棂等元素，烘托民国建筑风格。

规划设计：人性尺度的规划空间

城市空间的发展是一种内在的在原肌理上的生长。城市空间肌理的形成并非是完全偶然的，而是有一定规律可循的。项目旨在恢复传统街道活力，创造城市街区的生活感受，以多重界面逐级递进的交通模式，体现人性化的空间尺度。在延续南京本地居住特征的前提下，挖掘传统精髓，融合现代元素，引入崭新的生活模式。

项目还以高层为“山”，东南低西北高的建筑天际线，不仅巧妙地利用当地的地形气候，也为城市界面打开了一面新景象。高层住宅沿西侧北侧城市干道布置，应对城市尺度，体现现代化的城市形象。低层住宅半围合其中，应对传统街道尺度，体现传统街巷院落的生活方式。二者张弛有度，相映成趣，与中国传统的选址方式——“背山面水”相呼应。

此外，贯穿小区南北的中心主要水系，体现江南风景的秀气。整个设计以生态、自然、健康为规划理念，结合别具一格的民国建筑风格，力求营造宁静庄重、优雅从容的居住环境氛围。

Design Theme: Pacific and Reminiscent ROC Style

The design excavates regional features instead of copying trend. It extracts Nanjing traditional ROC architectural elements to differentiate from other similar properties. The main wall covers gray bricks, embellishing red bricks in, decorating with gray roof and rose beige window lattice to show its respect to ROC architectural style.

Planning: Human-oriented Space

Urban space development usually grows with its own texture. Since it is not a historical accident, it obeys certain rules. The project aims to rejuvenate traditional streets and constructs a block with a living atmosphere, so the design employs multiinterface progressive transportation mold to display human-oriented spaces. In addition, it bases on Nanjing local living characteristics to excavate traditional essence and embrace modern elements, bringing a fresh life style.

High-rise buildings like mountains form skyline descending from northwest to southeast. The design smartly utilizes local terrain and climate conditions to open a new picture for Nanjing. The high-rise buildings construct with urban arterial roads along west side and north side to correspond to urbanized scale and show modern urban image; while low-rise mansions semi-enclose in to echo traditional scale and display an alley and courtyard life style. The two clusters interplay well with each other, echoing Chinese traditional building orientation: facing water and against mountain.

A central water system transit south to north of the community, expressing a Jiangnan landscape spirit. Ecology, nature and health are three key points in the planning, combining with unique ROC style architecture to create a sedate and elegant living environment. Living here, residents live comfortably, moreover they can experience a wonderful life.

建筑设计：回归人性化的院落空间

项目在进行建筑单体设计时，力求从空间本质出发，用单体的一个个错落有致的“面”创造丰富的场所感受，为城市居民提供一个诗意的生活空间，从而提升整体社区的内在品质。

合院联排别墅空间

摒除传统联排别墅的军营式单调的阵列关系，独创合院半围合空间。“Z”字形组团公共空间，对应传统城市空间中小尺度的曲折的巷道空间，创造富有生活情趣的场所感受。

叠加别墅

创新叠加户型为“有天有地”模式，北面二层设有通高独立门厅入户，配有一车位，赠送北向花园。户内别墅电梯设计，避免了上层住户上下楼只能走楼梯的劳累之举。三层户内花园设计，让叠加上层住户亦可体验近地感受。四层赠送南向露台花园。此外还赠送有地下室娱乐及洗衣空间。

三进四院式联排别墅

重塑传统民居多进院落空间，以“院”为主题，安排各个功能空间，使每个功能空间均相对独立，不互相干扰，满足住宅的私密性要求。穿插在各个功能空间之间的院落，为每个房间都提供了良好的视线对景和良好的采光与通风条件。所有卧室空间均为套房设计，有独立的更衣室和卫生间。

Architectural Design: Return to Human-oriented Courtyard Space

Monomer design lays emphasis on spatial connotation. Dislocated monomer faces form opulent place feelings. The design creates a poetic living space and elevates internal quality for the community.

Enclosed Townhouse Space

Abandoning traditional townhouse monotonous array layout, the design innovates a semi-enclosed courtyard space. The public area is composed by “Z” clusters to echo small-scale meandering alley space in traditional urban space, creating a lifelike living place.

Overlay Villa

Innovative overlay villa house type is in a special mold. At ground floor residents have a north face two floors high independent entrance hall and a garage, and presents a northern garden for free. The villa equips with an elevator for upper residents. The third floor residents own an inner garden for them have fun to play with mud. The fourth floor residents are presented a terrace garden and a basement as entertainment room or laundry space.

Three Entrances Four Courtyards Townhouse

The design remolds traditional multi-entrance courtyard spaces, and takes court as theme to arrange functional spaces. Every functional space is independent without interruption, which guarantees private needs. The courts insert among functional spaces, provide every room with good vision landscapes, enough lighting and excellent ventilation conditions. Every bedroom sets as a suite with independent dressing room and bath room.

景观设计

景观设计意在打造“溪边坡地错落宅院”的景观意境，通过中央水系将中心景观与蜿蜒水系相结合，营造出推窗即景、移步换景的景观特色。在植物配置上，考虑四季不同色彩，运用几十种植物营造“春有花、夏有荫、秋有色、冬有绿”的四季景观。

Landscape Design

Landscape design aims to innovate a landscape layout that mansions are dislocating on slope aside streams. The central water system connects up central landscapes and meandering waterside landscapes. When residents open window they can enjoy those beautiful scenes; or stroll in the community to enjoy the scenes changing with moving steps. Considering four seasons have different colors, the design sets corresponding plants to echo four seasons landscapes and forms an environment of blossom in spring, shading in summer, colorful in autumn and green in winter.

源于上海石库门 植在长春新商业

长春绿地中央广场·饕界

Derive from Shanghai Shikumen Plant New Business in Changchun

Greenland Central Plaza Appetiting Dragon, Changchun

开发商：绿地集团 建筑设计：水石国际 景观设计：上海润实景观设计有限公司
项目地址：长春 建筑面积：20 000 平方米

Developer: Greenland Group Architecture Design: W&R Group
Landscape Design: Shanghai Runshi Landscape Design Co., Ltd. Location: Changchun Floor Area: 20,000 m²

长春和上海有着类似的城市精神——兼容并蓄、海纳百川。因此本案的设计引入了上海的城市元素，将石库门文化从上海移植到长春，不仅在立面选材上尽量移植和模仿石库门建筑元素，同时融入现代元素，在建筑布局上，更是将上海里弄文化进行了深入的剖析，进行全新演绎，既可体现商业街繁荣热闹的一面，又保留了石库门恬静淡雅的生活气息，为长春的城市商业带来新的活力。

Changchun and Shanghai have similar urban spirits that both of them are inclusive in all aspects. Hence the project is to introduce in Shanghai Shikumen architectural culture in Changchun. It imitates and transplants Shikumen elements on facade design, meanwhile modern elements are also added in. As for overall layout, the project uses Shanghai lane system as reference, and displays it in a brand new appearance. The design is conducive to present the prosperous and boisterous of the commercial street, and meanwhile it still keeps a tranquil and elegant life ambiance of Shikumen.

鸟瞰图 Aerial View

区位分析

项目位于长春市南部新城，东南临幸福街，西临丙十路，北临南环路，基地成梯形，总用地面积66 200平方米，要求容积率不大于3.5。项目基地位置将成为长春南部新城的核心地段。饕界区用地面积14 600平方米。

项目背景

饕界是绿地集团的室外商业街全国性品牌，这是绿地在北方的第一个特色商业项目。

长春，吉林省会，是吉林省的政治、经济、文化、交通和国际交往的中心，是东北地区中心城市之一，近海沿边开放城市，其本地人口与外地人口比例为1：2至1：3，移民人口占多数。这一点与上海有着很大的相似性，都具有海纳百川的包容性的城市精神。

设计理念

石库门，海派文化的经典元素；长春，多元文化汇集的移民城市。两者结合，体现了长春包容、多元的城市精神，也体现了海派文化海纳百川、兼容并蓄的本质。

Location Overview

The project locates in southern new town of Changchun, facing Xingfu Street in the southeast, Bingshi Road in the west and Nanhuan Road in the north. The project site is like a trapezoid of 66,200 m^2, plot ratio within 3.5. The site will be a core area in the southern new town in Changchun, and the project's floor area reaches 14,600 m^2.

Project Context

Appetiting Dragon is a national brand of outdoor pedestrian street of Greenland Group, and the project is Greenland Group the first featured business project in northern of China.

Changchun is the provincial capital in Jilin, and it is the center of politics, economy, culture, transportation and international association in Jilin as well as one of central cities in northeast of China. Changchun is close to sea, and it is a typical immigrant city with local population and migrant population proportion of 1：2 to 1：3, migrant population in majority, which owns the similarity with Shanghai, so it also enjoys inclusive urban spirit.

Design Conception

Shanghai Shikumen is a classical element in Shanghai-style culture. Changchun is a multi-cultured migrant city, so planting Shikumen architectural style in Changchun expresses its inclusive and multi-component urban spirits, and meanwhile the design conception is incarnated in Shanghai-style culture's nature of comprehensive and tolerant.

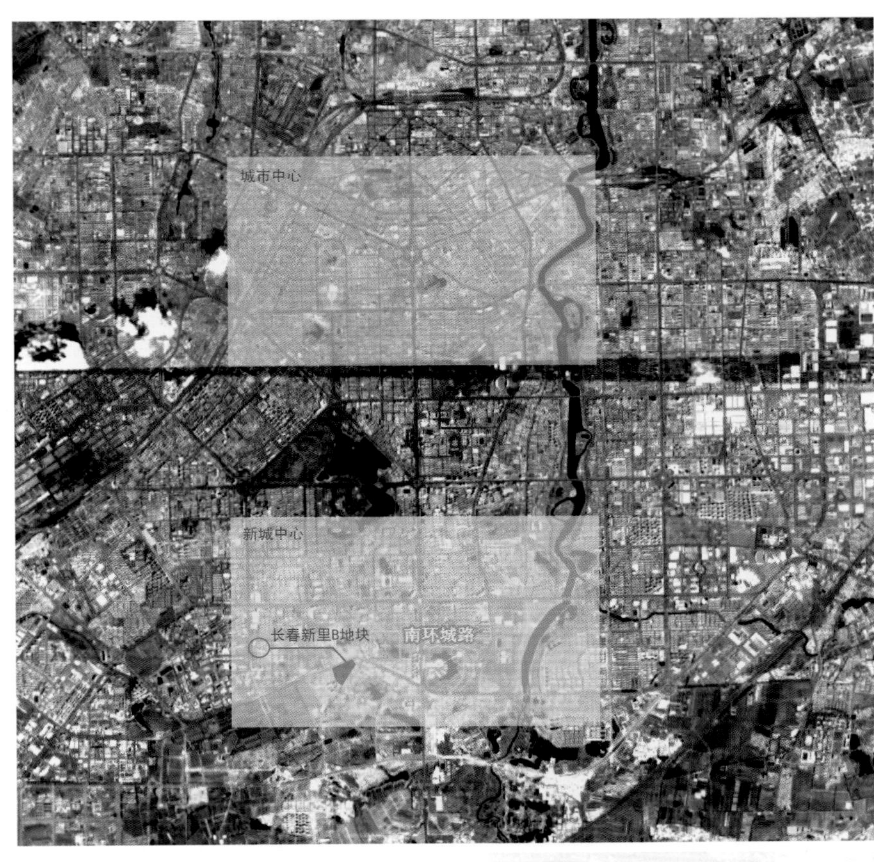

区位图 Location Map

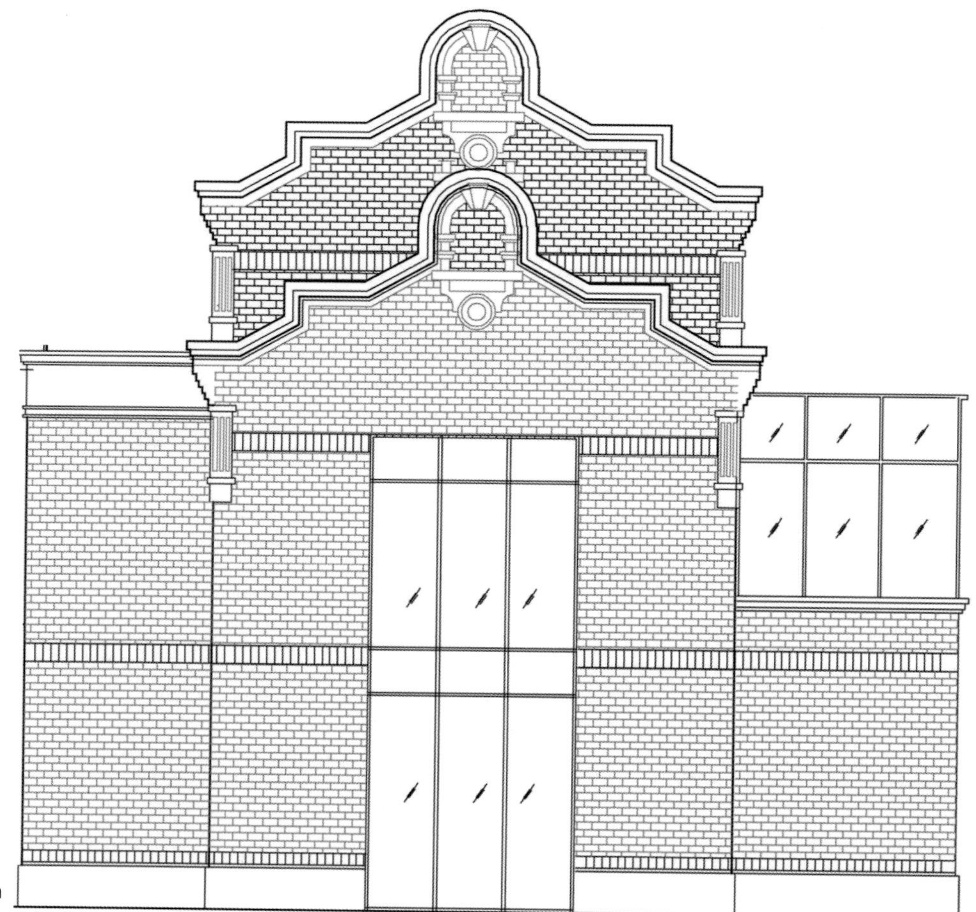

B2-A-A3立面图 B2-A-A3 Elevation

B3-7-1立面图 B3-7-1 Elevation

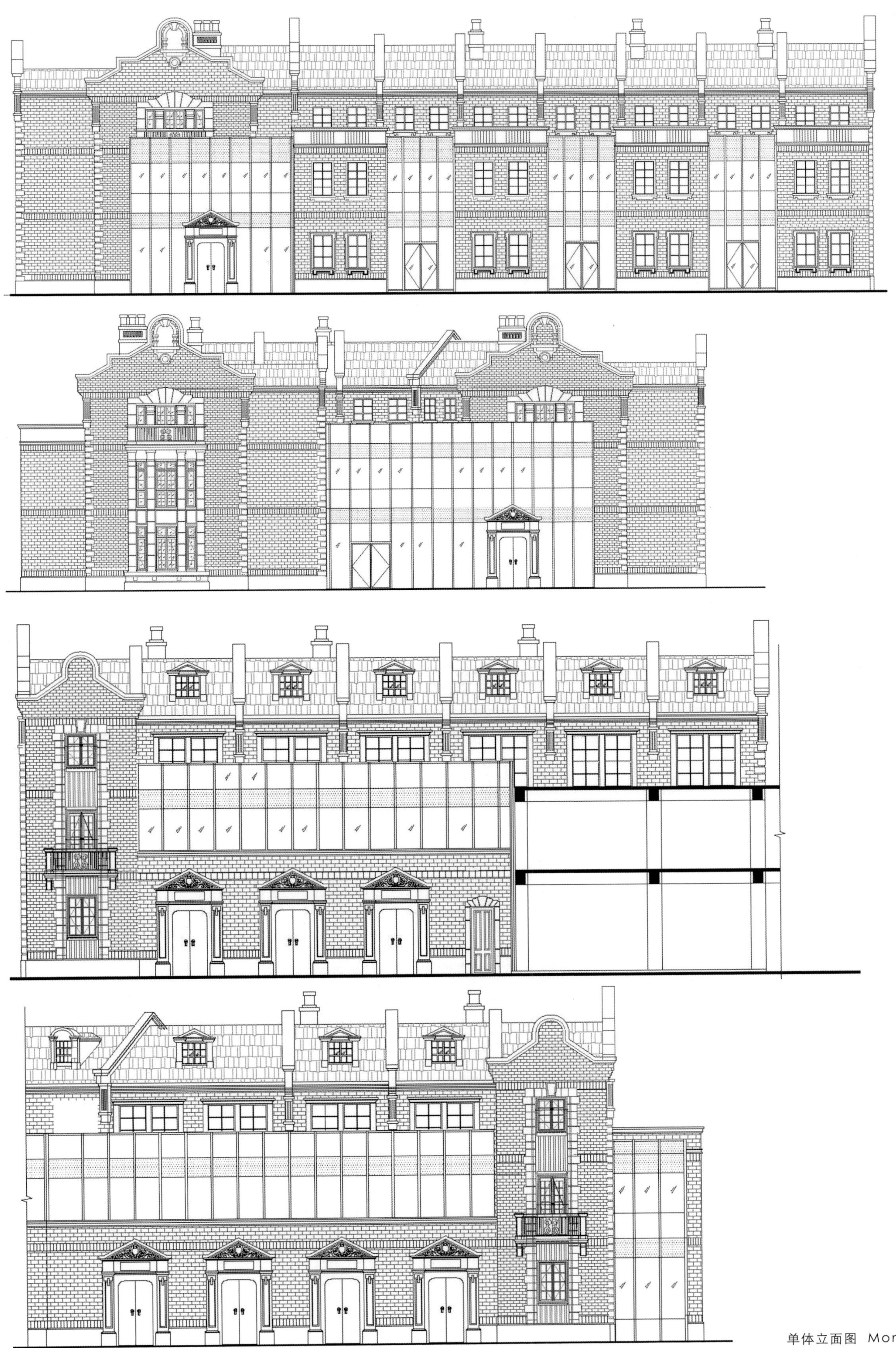

单体立面图 Monomer Elevation

建筑设计

项目设计以石库门为原型，植入现代元素，通过加法，既保持了石库门原有的气质，又保证了现代商业的功能需求。建筑总体布局充分考虑石库门的里弄特色，主弄、次弄、广场合理布置，闹静结合，既可体现商业街繁荣热闹的一面，又保留了石库门恬静淡雅的生活气息。

北方城市宽阔疏朗，加上新区规划普遍尺度较大，而上海老城的街巷里弄具备尺度适宜的步行体系，希望凭借在一个新兴居住区植入层次丰富的空间肌理，来激活社区的商业活力。

Architectural Design

Taking Shikumen architecture as phototype, the project plants in modern elements to retain original flavor of Shikumen and guarantee modern business functional requirements at the same time. The overall layout fully takes Shikumen lane system feature that main lanes, subordinated lanes and squares are in a certain arrangement which is conducive to present the prosperous and boisterous of the commercial street, and meanwhile it still keeps a tranquil and elegant life ambiance of Shikumen.

Northern city enjoys broad streets, and the new town also has large planning area. Depending on the proper pedestrian system in old Shanghai lanes, the project will build a newly-developing residential area with multilayered spaces so as to activate commercial activities.

一层平面 1F Plan

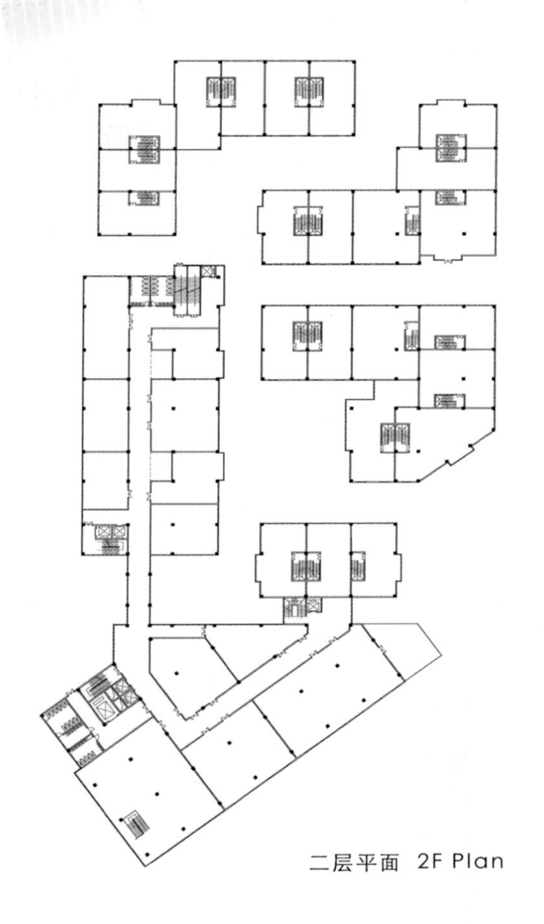

二层平面 2F Plan

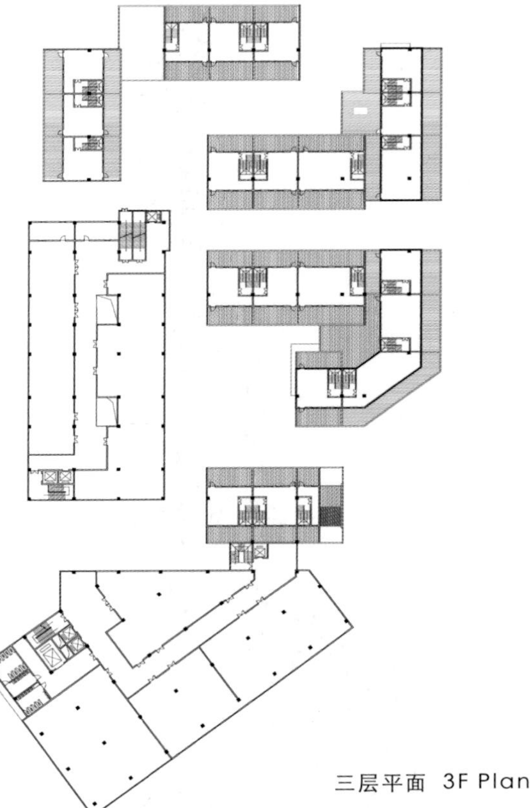

三层平面 3F Plan

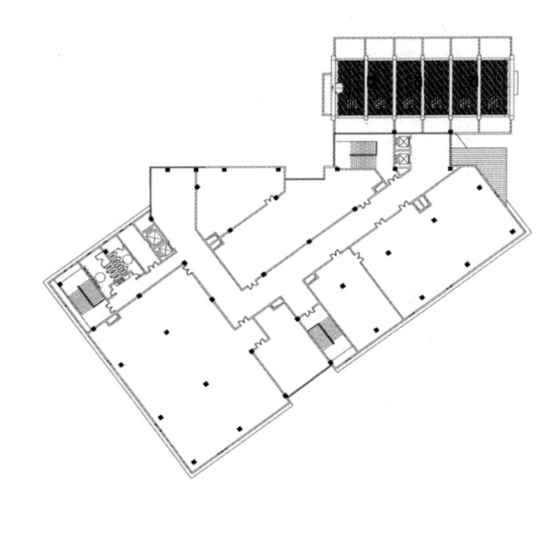

四层平面 4F Plan

室内设计

设计沿用里弄住宅的平面和空间组织方式，保证组团建筑在体量上符合传统街区的空间尺度和形式逻辑，同时，对室内尺度进行推敲，剖面上保留进一步改造的可能。100 平方米的单元面积构成基本的销售单位。在实际商铺销售中，客户根据需要购买相邻的单元进行改造；更多的是将商业零售与办公展厅在竖向上叠加混合。考虑北方积雪期，室外小尺度街道部分覆盖了易于清理的采光顶盖，实际创造了多一个层次的半室外商业空间。

Interior Design

The interior design employs lane residential plane and space organization, so that architecture clusters comply with the scale and form of traditional block space. As for the interior space scale, the design reserves a blank profile for further modification. Every selling unite has 100 m² area, and certainly clients can buy contiguous units into a larger one. In order to offer a convenient running environment, the office rooms are right up on shops. Considering the northern snow mantle period, small scale outdoor streets cover lighting canopy which is easy to clean up. Actually, the design creates an extra semi-outdoor commercial space.

天津水乡村落 移植江南元素

天津中国国家画院盘龙谷创作基地

Tianjin Water Village　Transplant Jiangnan Element

Panlong Valley National Painting and Calligraphy Creation Base, Tianjin

开发商：绿地集团　建筑设计：UA国际　项目地址：天津市蓟县5A级国家名胜风景区
用地面积：58 175.21平方米　总建筑面积：113 380.71平方米　地上建筑面积：5 403.37平方米　半地下建筑面积：5 857.16平方米　采编：张培华

Developer: Greenland Group　Architecture Design: Urban Architecture　Location: Jixian County, Tianjin
Site Area: 58,175.21 m²　Building Area: 113,380.71 m²　Ground Building Area: 5,403.37 m²　Semi-underground Building Area: 5,857.16 m²
Contributing Coordinator: Zhang Peihua

项目依托优美的自然环境及使用者的文化背景，将“白墙黑瓦间的幽谷雅韵”的江南建筑移植到天津，在天津构建起一处富有江南水乡民居意境的艺术创作基地。通过形似与神似的手法，在细部结构上既有传统砖雕门楼又有利用钢架栅格与玻璃材质模仿的披檐形式，突显江南建筑的特色。在空间设计上，更是将庭院、天井、入口等虚实空间的组织进行全新的演绎，既传达依山傍水的村落感，又赋予项目以江南园林天人合一的诗情画意的意境。

Depending on beautiful and natural environment and target users cultural backgrounds, the project transplants white wall and black tile featured Jiangnan architecture to Tianjin. It is a Jiangnan water village as an artistic creation base in Tianjin. The design takes transplant and imitation techniques to arrange traditional stone carving gateway and siding eave, composed of steel structure and glass clapboard, displaying Jiangnan architectural features. As for space design, the void and solid combination of courtyard, patio and entrance expresses a rural village flair and Jiangnan garden poetic feelings.

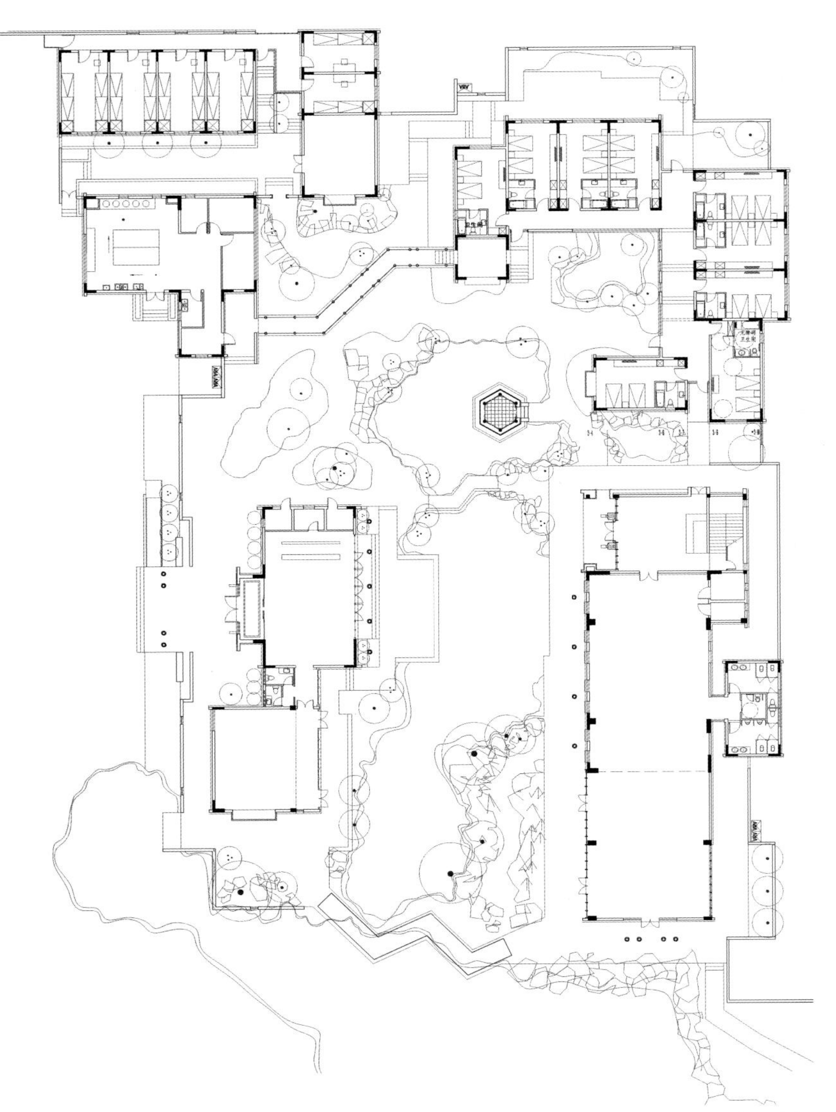

公共创作院平面图 Public Creation Center Plan

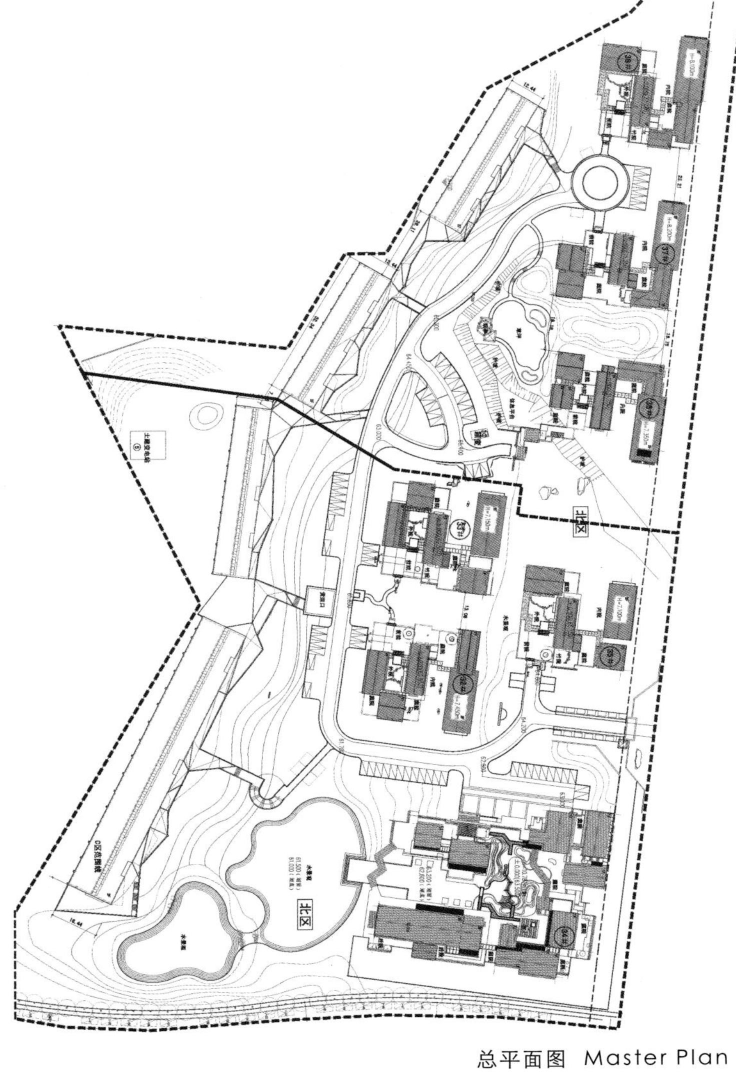

总平面图 Master Plan

总平面图 Master Plan

区位分析

“中国国家画院创作基地”坐落于天津蓟县5A级国家名胜风景区中。

设计背景

项目设计之思路始于业主对空间意境与文化意义的深度诉求。随着大众对传统文化的重新认可，这些深层次的观念需要在物质世界中得以重新展现，建筑的灵魂正是对“文化和情感”的诉求。然而，传统建筑中蕴含着无穷奥妙，却与现代生活方式存在一定差距。在设计之时和之后，依然需要思考的是如何将传统文化内涵附着于现代建筑设计与建造之中依然存在诸多的可能性。

项目概况

项目为画院艺术家提供艺术创作、交流与展示的场所。整个“创作基地”由一栋公共创作中心和多组个人工作室、大画室等建筑群组成。

Location Overview

National Painting and Calligraphy Creation Base locates in the 5A national scenic spot in Jixian County, Tianjin.

Project Context

The project design starts with a deep appeal for spatial and cultural meanings. Nowadays, people has reaccreditation to traditional culture, and the recognition needs a physical carrier to express, while architecture is just an ideal presentation form to pursue culture and affections. Nevertheless, the secrets in traditional architecture are not always in accordance with modern life style. Therefore, during the design and thereafter, we should consider how to blend traditional culture into modern architecture, and meanwhile explore more possibilities in design.

Project Overview

The project is a place for painters creating, communicating and displaying. The whole project consists of a public creation center and many individual studios and large-scale studios.

空间的组景与意象

项目地处视野开阔、山环水绕的风景名胜中，整个基地对内呈现半封闭状态，院落层层叠合，围合成一个小型聚落。空间的意境重点在于庭院、天井、入口等虚实空间的组织。多个独立庭院建筑以自然的状态组合成一个小型聚落，其中较大的创作中心象征村落的中心，而考虑特殊采光的画室采用了覆土设计的手法。将“聚落—院落—庭院”的布局方式在平面维度上展开，在自然山水环境下更似一片依山傍水的乡野村落。

在功能布局方面，无论规模或大或小，院落中主要建筑多考虑南北方向布置。建筑之间交接之处通过连廊、门楼相连接。建筑师充分利用建筑与庭院的连接部位营造不同空间效果。在院落内部的柱廊尽头、砖墙的转角处方向多次开启小天井式的空间，或利用狭小的漏窗，由此引入人的视觉通道，深奥曲折，通前达后。同时，为保证内外空间的流通性和私密性，采用传统的漏窗、矮墙、影壁与现代玻璃的混搭，实现与传统园林空间中“曲折尽致”相呼应，进而达到“眼前有景”效果。

符号的移植与模仿

项目整体氛围呈现出素雅恬静的乡土气质，细部构件与符号的处理中采用“形似”与“神似”两种相反的手法。建筑单元的入口直接移植传统砖雕门楼形式，实现形式的效果。同时，将具有传统建筑形式通过现代建筑语言进行重构，利用钢架栅格与玻璃材质模仿披檐形式，突出廊下空间的轻灵通透，避免木构屋面沉重之感，达到“神似”的效果。

Spatial Group Scene and Image

The project situates in a national scenic spot with broad vision and beautiful landscapes. The architecture layout is semi-closed inward to the site. Courtyards overlaps one another forming a small settlement. Spatial organization emphasizes void and solid combination of courtyard, patio and entrance. Many individual courtyards enclose to a natural village, and a bigger creation center in the middle symbolizes the center of the village. The design takes earth-sheltered technology to guarantee the requirements of lighting for painting rooms. The layout of “settlement-courtyard-yard” spreads on a plane dimension, which is like a rural village settling in a natural landscape environment.

Considering functional layout, the main buildings are in south-north orientation, no matter small scale or large scale. The junctions between buildings are connected by vestibules and gateways. Designers take these junctions to create various spaces. In courtyard, at the end of portico and at the corner of wall several small patios are arranged. The design also takes ornamental perforated windows to bring a visual pass, meandering and connective to every corner. In order to keep circulating and private effects for inner and outer spaces, the design combines ornamental perforated window, dwarf wall, screen wall and glass to echo traditional garden meandering space model that beautiful scenes are witnessed at any point.

Symbol Transplant and Imitation

The overall project ambiance reveals an elegant and tranquil rural flair. In detail design and symbol application, the design applies “transplant”and “imitation”ways. At the entrances of every building, traditional stone carving gateways express transplant design. However, some traditional architectural patterns are reconstructed by modern architectural technologies. Steel structure and glass clapboard combination imitates traditional siding eave, which gives a transparent space under porch and prevents wooden structure’s depression, reflecting the imitation effect.

徽州印象

Huizhou Impression

徽州建筑，以自然文化为底蕴，以白墙、青瓦、黑墙边绘画出一幅古色古香的自然山水的画作，没有浓墨重彩，没有浮夸，有的只是素雅、低调以及历史沉淀。徽州建筑携手历史与现代文明将徽州建筑的神秘面纱呈现在大众的视野中，展示出朦胧与自然的美感。

徽州建筑起源徽州，却因徽州缙绅和商业集团势力的崛起和发展，将其带出徽州，在江南、江北的各大城镇扎根落户，尤其是在浙江及江西一带。这里的徽州建筑主要涵盖安徽、浙江、江西三地的建筑。

随着历史的发展，徽州建筑逐渐自成一派，形成徽派建筑。徽派建筑是中国古代社会后期成熟的一大古建流派，其风格最为鲜明的是传统民居，它集中地反映了徽州的山地特征、风水意愿和地域美饰倾向。此外徽派建筑有着较为灵活的多层院落布局，装饰上古朴优美，尤其是砖雕、木雕、石雕艺术赋予了建筑鲜活的色彩。“白墙黑瓦马头墙，三间五架双楼房，砖雕门罩石(雕)漏窗，木雕楹联显文华”“民居祠堂石牌坊，村头水口园林化，长街短巷巧分布，天人合一呈吉祥”，这些民间的说辞便是徽派建筑的写照。

而因各地的地理环境和人文观念的影响，各地的徽州建筑也有所不同，在功能和结构、装饰等方面会略有差异，形成独具本土特色的徽派建筑。同时亦随着社会经济和文化的发展，当下的徽派建筑融入了现代的科技与材料，推进了徽州建筑文化的发展。

在此，我们选择了具有徽州建筑特色的项目，从当地的地域文化、历史文脉、自然环境出发，从传统性、历史性、现代性着眼，剖析徽州建筑的发展演变与发展，并通过其他地区移植和模仿的徽派建筑形成对比。

安徽九华山涵月楼度假酒店因地处徽州建筑的发源地，在设计上更契合徽州民居建筑，与自然山水相融合的同时，并融入皖南地区的徽文化历史文脉，将村落元素、楼塔、亭台、水巷与错落的单元布局，从对景、借景的景观内庭，不同空间围合的内院，尺度舒适的客房组团，创造具徽州聚落特征的现代村落意向。

而浙江临海伟星和院在低层住宅设计上采用了简化后的古典徽派建筑风格，而在布局上则引入唐长安城里坊制城市规划概念，将徽派与唐风融为一体，同时，又注重对自然现山水的利用，与自然成一体，为徽派建筑增添新的色彩。

而位于海口市的远洋华墅在设计上则是移植了徽派建筑元素，将徽派的村落及传统文化元素与海口的地域与环境相结合，将徽派建筑的自然闲适与海南的度假风情融为一体，为项目更增加了悠闲之感及沉静的气质，很好地发挥了徽派建筑的优势。

Hui-style architecture is like a quaint landscape painting depicting white walls and gray tiles, and it is without bright colors or exaggeration, only leaves an elegant and low profile image. Elegance and low key are on the strength of its cultural and historic foundations. Hui-style architecture wears mysterious veil in historical and modern civilization showing in front of us, presenting its hazy and natural aesthetic feelings.

Huizhou architecture is originated in Huizhou, and with the development of politics and economy, local officials and merchants have taken their homes to large towns around, especially in Zhejiang and Jiangxi. So in this chapter, the projects are collected from Anhui, Zhejiang and Jiangxi. As history progresses, Huizhou architecture owns its typical styles and forms Hui-style architecture.

Hui-style architecture was a mature ancient architecture genre in the late of ancient society of China. The most distinctive features are on traditional dwellings, because they intensively reflect mountainous characteristics, geomantic omen and regional ornament customs in Huizhou. Hui-style architecture has flexible courtyard layout and pristine ornaments, such as brick carvings, wood carvings and stone carvings. Many local folk rhymes contain featured elements in Hui-style architecture: white wall, gray tile, wharf wall, lattice window, ancestral hall, stone memorial archway, water gap, garden, long street and short lane.

Since geographic environment and humanistic conception are different in various regions, Hui-style architecture appears to be different in function, structure and ornament, and forms unique native characteristic Hui-style architecture. With economic and cultural developments, the Hui-style architecture absorbs modern technologies and materials, which in return promotes Hui-style architectural culture.

In this chapter, three Hui-style projects are chosen to present Hui-style architecture's development and evolution on regional culture, historical context and natural environment from traditional, historic and modern perspectives.

Jiu Hua Shan Han Yue Lou Resort & Spa locates in the birthplace of Hui-style architecture, so it fully echoes Huizhou dwelling characteristics. The design integrates architecture with natural landscapes, and adds Southern-Anhui historic context as well. Inspired by natural village aggregation formation, the design arranges tower, pavilion, water alley and dislocated apartments accordingly. Landscape courtyard adopts view-responding and view-borrowing methods to create comfortable guest room clusters. A modern Hui-style settlement-like village appears.

Weixing He Yuan adopts minimal classic Hui-style on low-rise residence, and uses Chang'an alley system planning concept of the Tang dynasty as a reference, well combining Hui-style with Tang dynasty flair. The design also concerns utilization of natural landscape resources, so it adds a new appearance to Hui-style architecture.

Sino Ocean Zen House locates in Haikou, and the project transplants Hui-style elements, such as Hui-style village and its traditional culture elements to integrate in local environment. The finished project containing Hui-style relaxation and Hainan holiday flavor. The transplantation not only adds relaxed and calm feelings to the project, but also boasts advantages of Hui-style architecture.

皖南院落空间 禅意旅居生活

安徽九华山涵月楼度假酒店

Southern Anhui Courtyard Space　Buddhist Mood Sojourn Living

Jiu Hua Shan Han Yue Lou Resort & Spa, Anhui

开发商：雨润集团　规划及建筑设计：上海秉仁建筑师事务所　施工图设计：南京市凯盛建筑设计研究院有限责任公司　景观设计：上海北斗星景观设计有限公司
室内设计：美国威尔逊室内建筑设计公司　项目地址：安徽省池州市九华山风景区柯村新区管委会北门对面
用地面积：207 610.8平方米　建筑面积：96 458平方米　容积率：0.4　绿地率：48%　采编：康欣

Developer: Yurun Group　Planning & Architecture Design: DDB International Ltd. Shanghai
Construction Drawing Design: Nanjing Kaicheng Institute of Architectural Design & Research　Landscape Design: Triones Landscape Design Co. Ltd.
Interior Design: Wilson Associates | Interior Architectural Design　Location: Ko Village District, Jiuhua Mountain Scenic Area, Chizhou, Anhui
Site Area: 207,610.8 m^2　Building Area: 96,458 m^2　Plot Ratio: 0.4　Greening Rate: 48%　Contributing Coordinator: Kang Xin

项目地处九华山，以皖南地区徽文化作为历史文脉，享有九华山自然景观和深厚的佛教文化资源，八字门楼、石雕漏窗、徽砖瓦、马头墙、隔扇窗等徽州民居元素和符号，传达出徽州民居建筑的原汁原味之感。而楼塔、亭台、水巷和错落的单元布局，再现了徽州聚落形态的自然村落，以现代的手法诠释出徽州民居建筑的文化脉络。此外，项目中的莲花与莲叶的印记，亦契合九华山的人文情怀，营造出富有禅意的生活空间。

The project situates in the Jiuhua Mountain. It takes historical context of southern Anhui, and enjoys rich natural resources and profound Buddhist culture from the Jiuhua Mountain. The elements and symbols of Hui-style dwellings such as "八" character gatehouse, perforated rock window, Hui-style brick and tile, wharf wall and partition board window all present original flair of Hui-style dwellings. The tower, pavilion, terrace, water alley and dislocated unit layout together restore natural village of Huizhou settlement. The graphic patterns of lotus and its leaf are in accordance with humanistic feelings of the Jiuhua Mountain, creating a Buddhist mood living space.

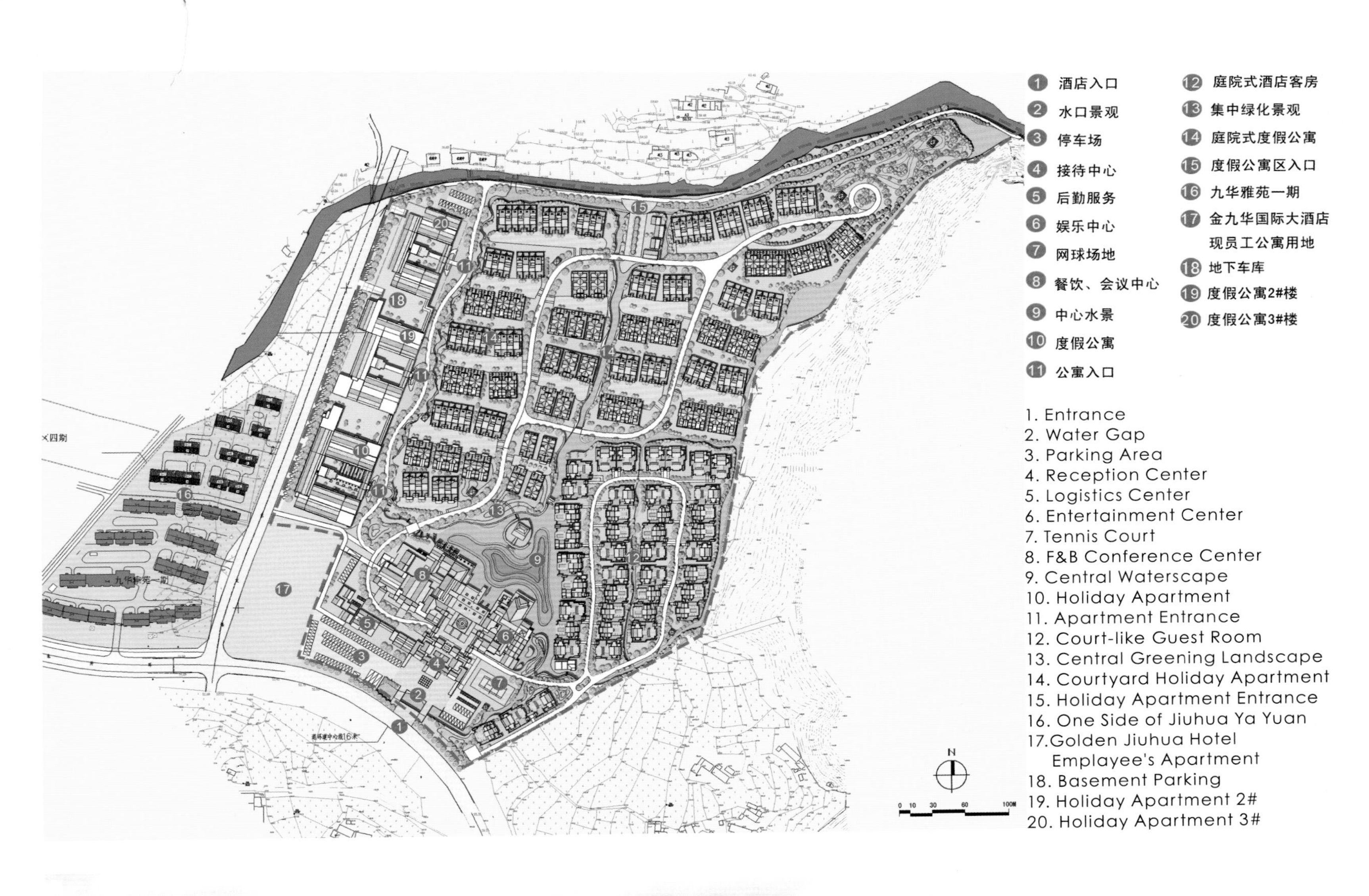
1 酒店入口
2 水口景观
3 停车场
4 接待中心
5 后勤服务
6 娱乐中心
7 网球场地
8 餐饮、会议中心
9 中心水景
10 度假公寓
11 公寓入口
12 庭院式酒店客房
13 集中绿化景观
14 庭院式度假公寓
15 度假公寓区入口
16 九华雅苑一期
17 金九华国际大酒店现员工公寓用地
18 地下车库
19 度假公寓2#楼
20 度假公寓3#楼
1. Entrance
2. Water Gap
3. Parking Area
4. Reception Center
5. Logistics Center
6. Entertainment Center
7. Tennis Court
8. F&B Conference Center
9. Central Waterscape
10. Holiday Apartment
11. Apartment Entrance
12. Court-like Guest Room
13. Central Greening Landscape
14. Courtyard Holiday Apartment
15. Holiday Apartment Entrance
16. One Side of Jiuhua Ya Yuan
17.Golden Jiuhua Hotel Emplayee's Apartment
18. Basement Parking
19. Holiday Apartment 2#
20. Holiday Apartment 3#
N
0 10 30 60 100M

区位分析

项目基地位于九华山风景保护区西侧，在建99米高的地藏菩萨大铜像北侧偏东。基地东侧为九华山山体，南侧为城市道路，西侧临规划道路为四层住宅小区，北面临溪流自东向西。基地内为未建设用地，地势较平坦，东南至西北略有落差。

项目背景

徽文化历史久远，徽州建筑文化亦是集徽州山川风景的灵气，融合风俗文化的精华。空间结构上，造型丰富，讲究韵律美，以马头墙、小青瓦最有特色，而石雕、木雕、砖雕等装饰，使建筑更显富丽堂皇。

九华山风景区位于安徽省池州市境内，是世界一流旅游胜地、国际重要佛教道场。这座名山自汉代以来沉积了灿烂辉煌的文化遗产，以致形成今天的风景名山、历史名山、文化名山、佛教名山。从明代起，九华山就是中国佛教四大名山之一，被誉为“莲花佛国”。 基地位居于此，拥有丰富的自然资源。

Location Overview

The project situates in the west of Jiuhua Mountain Scenic Conservation Area, with a 99 m constructing Earth Store Bodhisattva Bronze Statue in the east-north. The project's site is near the Jiuhua Mountain in the east with Urban Road in the south and Planning Road in the west where a community of four-floor residences stands up, and there is a stream flowing from east to west in the north. The plot had not been constructed yet, and it enjoys flat terrain, gently descending from southeast to northwest.

Project Context

Hui-style culture has a time-honored history, and has gathered Huizhou landscape spirit and customs essence. Hui-style architecture features in opulent spatial configuration and rhythmic aesthetics.Wharf wall, Chinese-style gray tile are of most distinctive, and stone carving, wood sculpture and brick carving make the architecture look splendid.

Locating in Chizhou, Anhui, the Jiuhua Mountain scenic spot is a first-class tourist spot and international influential Buddhism ashram. The well-known mountain has deposited splendid cultural heritage since Han dynasty. Today, it is famous for its great landscapes, profound history, rich culture and strong Buddhist resources. From Ming dynasty, the Jiuhua Mountain has been one of four Buddhist mountains in China, and was regarded as a “Lotus Buddhism Country”.

规划设计

轴线性的布局形态

中心轴以水口园林为酒店入口空间起点，庭院内外开合，景观层叠递进，于中心景观水池——莲心池达到高潮。村落形态的酒店客房单元组团分布于中心的东南侧，庭院式度假公寓区域位于基地的北半侧，度假酒店式公寓位于基地西侧。

庭、院、村落式的空间营造

设计以自然村落的聚合形态为灵感，将村落元素楼塔、亭台、水巷与错落的单元布局，从对景、借景的景观内庭，不同空间围合的内院，尺度舒适的客房组团，创造具徽州聚落特征的现代意向村落。客房组团间水带蜿蜒，布局错落，集中水域、莲花池与景观塔、亭的布局形式于道路的流转中映衬出印象中的徽州聚落民居。

Planning Design

Axis Layout

A water gap garden on central axis forms the entrance of the hotel , then walking through external and internal courtyards, along layer-and-layer progressive landscapes, to the central landscape-the Lotus Pool that the whole axis scenery is unfolded up. Village-like guestroom clusters squat in the southeast of the center, courtyard-like tourist apartment clusters sit in the north of the site, and tourist apartment area is in the west.

Court, Yard and Village-like Space

Inspired by natural village aggregation formation, the design arranges tower, pavilion, water alley and dislocated apartments according to the same model. Landscape courtyard adopts view-responding and view-borrowing methods to create comfortable guest room clusters that are a modern Hui-style settlement-like village. Dislocated guest room clusters are connected by a meandering water system. The Lotus Pool, Landscape Tower and layout of pavilions intersperse along roads, creating a Hui-style dwelling settlement.

建筑设计

客房组团区分为客房单元组团和一个组团中心服务区。客房单元组团由中心服务区往东，以村落式的布局形态，结合道路和水域景观，依次分布在基地的东南部，相对独立，以水域相隔，营造根植于皖南大地的院落式酒店居住空间。

客房单元造型汲取徽州民居典型的造型元素，运用八字门楼、石雕漏窗、徽砖瓦等原味的徽州民居元素与马头墙、隔扇窗等写意的徽派民居符号，结合黑白横竖、水道蜿蜒等抽象的徽州民居印象，塑造原汁原味的徽州院落式酒店客房单元。

Architectural Design

The guest room cluster area consists of guest room units and a service center. The guest room units situate in the east of the service center in a village model. These units combine with roads and waterscapes freestanding in the southeast of the site and separated by a stretch water, creating a southern Anhui courtyard living space.

The guest room unit employs typical elements of Hui-style, such as "八" character gatehouse, perforated rock window, Hui-style brick and tile, and vernacular dwelling symbols. Wharf wall and partition board window combining black and white color collocation and horizontal and vertical textures as well as meandering water system, original Hui-style courtyard-like hotel guest rooms are perfectly presented.

景观设计

项目中心区域景观结合建筑的形态，根据庭院广场水域的开合递进，于景观节点中设置亭、塔、廊等元素，使视觉景观逐步展开。水域空间与流线的结合更使中心景观秩序清晰，视觉层次丰富。

中心景观水池开阔的水域空间，以其形态的象征性——莲，显露出佛的宽广意义，临水建筑的亲水平台亦丰富了建筑的景观格调。中心木塔形式取材歙县许村，不仅成为景观的视觉重心，也给予景观以皖南民居村落的烙印。

在水系景观架构中，均折射出这些流传于徽州丰富的水文化形态。徽州村落的风水屏障——水口，与古树、亭、桥构成独特的水口园林景观。巷间蜿蜒流转的水道化城寺的莲花池取水于自然，庭院水景以不同形态构成月松阁的水域层次，并与楼、塔、亭、阁相辅相融，平添灵韵。

此外，结合酒店的服务流线、后勤流线、景观休闲流线，将道路的形式结合组团的形态、整体景观构架、地形起伏等因素，且道路节点上设置具徽州村落特色的木塔、植栽，在道路线型上亦融合了九华山“莲花佛国”的吉祥象征——莲叶形的宛转离合，不仅感受到自然村落形态的印象，亦契合了九华山的人文情怀。

Landscape Design

The central landscape works in concert with the architectural formation to arrange pavilion, tower and gallery in courtyard plaza, water area and landscape nodes, unfolding visual landscapes step by step. The combination of water areas and pedestrian circulation enriches visual hierarchies, making the central landscape have clearer order.

Spacious central landscape pool looks like a lotus, implicating the vast of Buddhism. Waterfront terrace enriches landscape pattern. Central wood tower draws materials from Shexian County Xu Village, which is a visual focus, leaving a southern Anhui dwelling village style on the landscape.

The waterscape implicates rich water culture that handed down from ancient in Huizhou. A typical Hui-style protective screen-water gap, combining with old trees, pavilion and bridge, forms unique water gap garden. Natural water flows into winding water alleys and the Huacheng Temple Lotus Pool. Courtyard waterscapes are in various forms and constitute hierarchical water area in the Pine Pavilion.

In addition, the design unites hotel service circulation, logistic circulation and landscape circulation. The road patterns combine with cluster forms, the whole landscape structure and topography conditions. Hui-style village featured wooden tower and plants intersperse along road nodes, and the arrangements are like lotus leaves, implicating auspicious symbol of the “Lotus Buddhism Country”, which reminds people of a natural village flair and humanistic feelings of the Jiuhua Mountain.

江南水乡三重院落 中魂西技现代生活

南昌伟梦清水湾

Trio Courtyards in Jiangnan Water Town Modern Life with Chinese Spirit and Western Technology

Weimeng Clear Water Bay Community, Nanchang

开发商：南昌豪佳实业有限公司 设计单位：昂塞迪赛（北京）建筑设计有限公司 首席设计师：刘亮 项目经理：方华 项目地址：江西南昌
总建筑面积：636 718.07平方米 地上建筑面积：484 730.81平方米 容积率：1.26 建筑密度：25% 绿地率：36% 采编：郭学然

Developer: Nanchang Jia Hao Industrial Co., Ltd. Design Company: On-Site Design Group Chief Designer: Liu Liang Project Manager: Fang Hua
Location: Nanchang, Jiangxi Gross Building Area: 636,718.07 m² Ground Building Area: 484,730.81 m²
Plot Ratio: 1.26 Building Density: 25% Greening Rate: 36% Contributing Coordinator: Guo Xueran

南昌作为沿海地区商贸辐射中西部的中转枢纽，不仅受移民文化的影响，外来文化的影响也是不容忽略的。南昌清水湾在设计上，遵循江南园林格调的同时，引入徽派元素，并采用中魂西技的手法，不仅全新演绎了中式建筑文化中围合和院落文化的精髓，亦将西式和现代的技法融入其中，以一种全新的视角，将江南建筑淡雅明快的格调与现代简约的技法相融合，使建筑、自然与人和谐统一，形成富有未来气息及江南水乡特色的居住空间。

Nanchang is the transit hub in mid-west of coastal commerce and trade area, hence it is affected by immigrant culture and even by foreign culture. The project design follows Jiangnan garden style and imports Hui-style elements. Chinese Spirit and western technologies are combined in Chinese enclosure courtyard cultural essence. It is a brand new perspective to integrate Jiangnan architectural elegance and vibrance and modern concise techniques. The overall design makes architecture, nature and human being into a unity, creating a futuristic space Jiangnan water town characteristic .

总平面图 Master Plan

基地现状 Site Information

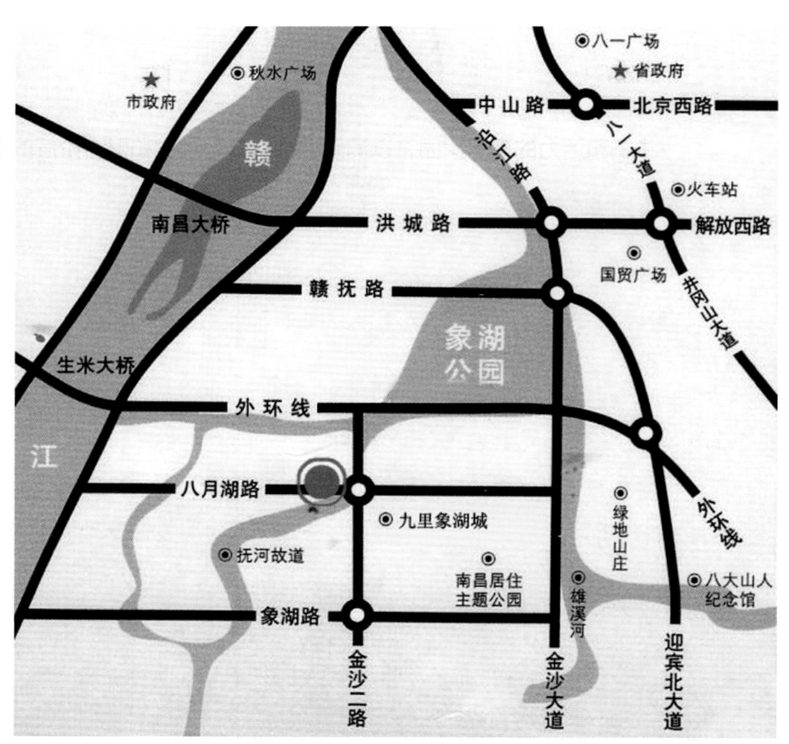

基地交通区位图 Site Traffic Location Map

区域分析

项目位于江西省南昌市南昌县象湖新城，环境优越，空气质量优良。项目北临八月湖路，西临诚信路，东侧、南侧为抚河故道，距象湖湿地公园景区仅1公里。项目周边各项配套设施较为匮乏。

项目背景

在城市发展的过程中，外来经济和外来文化的影响也是不可忽视的一部分，南昌在历史的发展过程中，徽商对其的影响也是不可小觑的，不仅是在经济上，在建筑文化上也留下了印记，在宏观的江南园林文化的基础上，徽派建筑成为了南昌建筑文化中的一部分。此外，南昌作为沿海地区商贸辐射中西部的中转枢纽，受外来文化的影响是必然的，中西文化交汇的同时，亦是在不断地磨合和融合，形成一种兼容并蓄的形式。

设计理念：城市中心，别院生活

项目设计从规划设计、园林景观，到会所配套，均强调客户舒适感受的营造与体验，充分体现客户的私密感、尊贵感和安全感的生活方式，体现“庭院深深”的居家理念；同时在城市近郊营造一处高品位、满足未来生活方式、具有生机和活力的理想家园，以满足城市权富、知富和财富阶层的生活需求。

规划结构

项目结合设计要求与景观规划将整个社区分为五大板块：沿八月湖路即用地北侧边界与沿诚信路北端布置沿街高层与商业板块；地块内部分为花园洋房板块、中密度住宅板块、低密度住宅板块及写字楼板块。项目居住部分采用围合型组团与院落相结合的结构形式。院落通过水系以及步行系统与组团绿地相联系，空间序列清晰，可识别性较强，同时又便于管理、方便居民使用。

Location Overview

The project locates in Xianghu New Town, Nanchang county, Nanchang, Jiangxi. It enjoys prominent environment and fresh air with Bayuehu Road in the north, Chengxin Road in the west and the limpid old Fuhe River running in the east and the south. The community is just 1 km away from Xianghu Wetland Park Scenic Spot. However, it goes short of supporting facilities at present.

Project Context

In the process of city development, external economic and cultural influences are indispensible factors. Anhui merchants has always affected Nanchang development on economy and architectural culture. On the basis of Jiangnan garden culture, Hui-style architecture has grown to be a part of Nanchang architectural culture. In addition, Nanchang is the transit hub in mid-west of coastal commerce and trade area, hence affected by external culture. While in the interaction of Chinese and Western cultures that are integrating and fitting together in the same time, forming an inclusive state.

Design Conception: Courtyard in Urban Center

Customer's comfortable feeling and experience are taken to the foreground in the design aspects of planning, garden landscape and supporting facilities. In order to show respect to customer's privacy, sense of security and life styles, the project creates a tranquil and exquisite courtyard community nearby the center of the city. It is an ideal community with high taste, futuristic life style and vitality, so it can satisfy the needs of the power, the rich and the intellectual.

Planning Layout

According to design requirement and landscape planning, the community is divided into five parts: the high-rise commercial building part along the north of Bayuehu Road and Chengxin Road, the garden house part, the medium density residence part, the low density residence part and the office building part. The residential part are in a form of combining enclosure cluster and courtyard. All the courtyards are connected by water system and pedestrian network, interweaving in the greenbelts. The whole layout appears clear spatial sequence and high identifiability, and it also is conducive to using and management.

建筑设计

小区住宅由南至北由叠加住宅、联排住宅、洋房、办公大楼、高层住宅组成，形成由低向高逐步的过渡。联排住宅及洋房强调建筑群体组合，内外空间的渗透，形式活泼、自由，建筑形式具有时代气息。

提取洋房语素

高层住宅提取洋房的立面语素，更多强调垂直线条的贯通，体现简洁、轻盈的建筑风格，通过细腻的手法进行阳台、檐口、色彩等细部的变化，创造出亲切典雅的现代中式的建筑形象。两者之间通过色彩、屋顶形式、建筑风格的呼应，保持整体感。

“中魂西技”手法

建筑风格充分考虑到该保护区的悠久历史和醇厚的文化底蕴，为了延续地域独有的历史文脉，在建筑设计上既秉承中式建筑的特点，又不乏西式的典雅。立面设计采用“中魂西技”的手法，运用石材、面砖、金属压顶，注重材料质感与色彩的对比与统一，强调尊崇的品质感。在材质上，采用深灰色的面砖、浅灰的石材、高档的红色木门，以深灰色金属压顶收边，局部设计原木色的木格栅，以现代元素体现中式建筑神韵。

新中式别墅建筑

项目采用了新中式的别墅风格，继承了中式建筑的围合布局形式。在立面色彩上，以体现低明度的暖灰色调，营造高贵典雅的现代中式建筑空间；以材料色彩上的沉稳勾勒出纯粹的中式味道，流露出建筑与自然色泽的和谐与融合。

Architectural Design

In the community overlaying residence, townhouse, garden house, office building and high-rise residence are gently ascending in order from south to north. And among them, the townhouse and garden house highlight group composition and penetrating spaces. They are in a lively and free style, emitting out the ambiance of the times.

Garden House Element

The high-rise residence adopts garden house facade elements and highlights vertical lines to present a cut-through, concise and lithe style. Ingenious skills are embodied on the variation of details, such as balcony, cornice and color collocation, creating an amiable and elegant modern Chinese style. The high-rise residence and garden house echo each other through color, roof configuration and the overall style.

Chinese Spirit and Western Technology

Considering Nanchang long history and profound cultural foundation, architectural style inherits local typical historical context. The design integrates Chinese architectural characteristics and western elegant flavor. The facade design adopts Chinese spirit and western technology. Stone, facing brick, metal coping comply with principles of contrast and unity of texture and color, and wall carvings are in very high quality. All the modern elements of dark gray facing brick, light gray stone and top-grade red wooden door, dark gray metal coping edge complementing with burlywood wooden grating create Chinese-style architectural spirits.

Neo-Chinese Villa

The villa applies neo-Chinese villa style and inherits Chinese enclosure layout. Chinese architecture has four schools: traditional Chinese symbolic school, Hui-style dwelling school, quadrangle courtyard school and reformist school. As for facade design, gray brick, gray window, red wooden door, burlywood wooden grating and dark gray metal coping edge present a low brightness warm gray, creating noble and elegant modern Chinese-style spaces. The calm colors of construction materials express Chinese flair, and meanwhile echo the lustre of nature.

景观设计：三重院落

景观设计采用三重庭院设计，把空间意识转化为时间进程，体现出一种场所精神。独具特色的入口区，成为清水湾院子区别于其他居住空间的个性特质。在自然的归属感中，找到惬意、闲适的生活感受。

整个居住区采用城市概念：庭院将整个居住区看作是一个城市的公园来设计；强调对城市空间和整体景观结构的呼应与氛围塑造；关注整体围合型的结构与空间层次内涵；使居住区域与绿地空间有机联系；并运用组团划分和空间管理的概念。

组团社区采用院落设计：庭院将自然融入生活，将自然引入围合庭院式的布局。组团入口空间的独立园景设计，形成组团居民共有的安全宁静的天地，将生活的自然主题放大成为艺术化展现。

给住户以家的舒适：在入户花园的设计中，“空间联想”的意图明显。通过模拟自然要素，借自然之精，缩小自然天地。

Landscape Design: Trio Courtyards

Landscape design takes trio courtyards layout, transforming space consciousness into time course and bringing a kind of place spirit. Unique entrance differentiate it from other residential spaces, presenting a pleased and relaxed life in nature.

Urban Concept Residence

Courtyard design regards the whole residential area as a urban park, so it applies urban space and comprehensive landscape structure in the courtyard layout. It also takes overall enclosure form and spatial level to organize residential area and greening spaces so as to achieve cluster division and spatial management.

Courtyard-like Cluster Community

The courtyard blends in natural life and draws nature in courtyard-like enclosure layout. The independent landscapes at the entrance of clusters build a quiet and safe space for inhabitants. Natural theme is expressed in an artistic way.

Comfortable Home

Spatial association incarnates on home garden design which imitates natural elements to concentrate on the essence of nature.

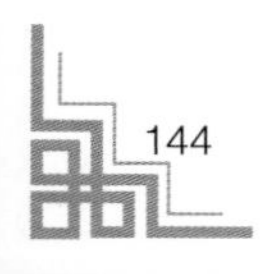

独栋A 一层平面图 Monomer A 1F Plan

独栋B 一层平面图 Monomer B 1F Plan

户型设计

高层住宅： 户型设计涵盖90~180平方米的多种户型，满足多层次的社会需求；在住宅楼入口处设置无障碍通道；平面布局具有超前性，功能完善，结构较为简单，更接近产业化的需求。户型设计中增设凸窗、宽敞窗台、观景阳台等，提高住宅采光、通风、日照效果，增加住宅的通透感和观景性能。

独栋别墅建筑： 根据当地的气候特征和居住习惯，设计着重对户型进行了创新尝试。空间布局南北通透，能有效组织空气对流，使居住空间有充分的新鲜空气，充分利用基地内优美的自然和人文景观，并采用5大花园设计，包括前花园、后花园、侧花园、天台花园和下沉式花园。从而使得从地下到天台，有天有地，全方位采光，通风、景观极佳。

双拼别墅建筑： 住宅南北皆设计有入户花园，并附设采光地下庭院。2~3层结合退台、露台、阳台设计，在为住户提供户外景观空间的同时，也丰富了建筑立面形象。在别墅的内部空间设置上，注重主场景空间的设计，呈现出不同的住宅产品特色。空间利用虚拟隔断、过渡、衔接、掩映、错落、对称等手法，呈现丰富的空间场域感，满足居住者多层次起居空间需求。

联排别墅建筑： 联排别墅通过精心布置，使每户客厅和主卧室、大型露台均朝向南。每种户型都有花园式大型露台或阳台，配合高低错落的屋面、退台，形成丰富错落的中式建筑。户型均有南北入户口、院落、南北露台等。

House Type Design

High-rise Residence includes multiple house types arranging from 90 m^2 to 180 m^2 to meet multileveled social needs. Barrier-free way is paved at every entrance of residence. Advanced plan layout has complete functions and concise structure, meeting industrialized requirements. The design adds bay window, spacious balcony and viewing balcony to improve daylighting and ventilation effects, bringing transparent vision and a landscape performance.

Detached Villa design innovates house types according to local climate and living habits. North-south ventilation is conducive to organize air circulation and draw in fresh air. The orientation fully utilizes natural and humanistic landscapes. Every detached villa includes five gardens: front garden, rear garden, side garden, roof garden and sunken garden, so it enjoys panoramic lighting and ventilation and extraordinary landscapes.

Semi-detached Villa equips a north garden, a south garden and a daylighting underground yard. The second and third floor have terrace and balcony to offer outdoor landscapes and enrich facade appearance in the same time. As for the internal space, the design highlights a main scenario space to make different from other residential product featured spaces. In order to improve spatial levels, virtual partition, transition, linking, shading, dislocating and symmetric techniques are applied to present spatial scenes so as to meet multiple living needs.

Townhouse organizes living room, master bedroom and large-sized terrace toward the south. Every house type equips a garden-like terrace or balcony, south-north entrances and courtyards in different heights and strewn at random.

临海文化根基 简约徽派建筑

浙江临海伟星和院

Linhai Cultural Foundation   Minimal Hui-style Architecture

Weixing He Yuan, Zhejiang

开发商：伟星集团　建筑设计：上海拓维都市设计顾问有限公司　项目地址：浙江临海
用地规模：158 667平方米　地上建筑面积：190 084平方米　地下建筑面积：111 872平方米　容积率：1.19　采编：康欣

Developer: Weixing Group　Architecture Design: Shanghai Topway Urban Design Consultant Co. Ltd.　Location: Linhai, Zhejiang
Site Area: 158,667 m²　Ground Building Area: 190,084 m²　Underground Building Area: 111,872 m²　Plot Ratio: 1.19　Contributing Coordinator: Kang Xin

临海作为历史文化名城，有着深厚的历史文化和建筑文化根基，并伴随着经济和文化的发展，建筑文化亦在不断地演变。在项目设计中，不仅沿用了历史传承已久的里坊制规划设计，采用了古典“十”字形规划设计概念，充分发挥了山体与南北向水渠的自然条件，增强了项目的自然性。同时也采用了现代简化后的徽派建筑元素，如低层住宅采用了坡顶设计，既丰富了建筑的第五立面，同时也解决了因当地多雨水和炎热等自然气候影响需要建筑保温隔热、防雨水的要求。此外，有徽派建筑常见的深色青砖，配以浅色涂料，加之细巧的线脚，既呼应徽派建筑的气质，又不失现代的精致简约。

Linhai is a historic city owning profound foundations of history and architectural culture. With the development of economy and culture, the architectural culture has been developed as well. The project design inherits alley system planning that handed down from ancient times, and adopts classic cruciform structure to connect the mountain and north-south water channel. The facade employs modern minimal Hui-style elements, such as sloping roof which enriches the fifth elevation while solves heat, insulated and water-proof issues due to local climate conditions that there is too much rain and scorching heat. In addition, the Hui-style dark brick and light painting matching with delicate skintle echoes the spirit of Hui-style architecture, concise yet exquisite.

技术经济指标一

项目		指标		说明
总用地面积（m²）		158667.7		-
地上建筑面积（m²）		190400.3		计入容积率
其中	低层住宅	59612.0	187763.5	-
	高层住宅	128151.5		其中物业用房300m²
	A-1#楼	2636.8		其中物业用房1030m²，公厕60m²
容积率		1.2		-
建筑密度		22.5%		-
绿地率		30.0%		-
住宅户数（户）	低层住宅	210	1091	-
	高层住宅	881		
总人口（人）		3491.2		3.2人/户
地下机动车停车量（辆）	低层住宅	673		3.2辆/户
	高层住宅	1090		大于1.0辆/户
	访客	33		3辆/百户
地下建筑面积（m²）	低层住宅	56816.0	101266.6	不计入容积率
	高层住宅	44450.6		

技术经济指标二

面积分类 \ 分期		一期	二期	三期	四期
分期用地面积		11926.1	21324.4	21321.4	110063.4
分期建筑面积	地上建筑面积	35188.5	55863.8	37099.2	62248.8
	地下建筑面积	7219.0	18781.4	18440.6	56816.0
	合计	42407.5	74645.2	55539.8	119064.8

规划总平面图
Site Plan

区域分析

基地位于临海市双绿路西北地块，位于临海中部的北区，基地东临双绿路，西北临自然山体，基地西南面为村宅现状。基地处于临海市市中心东北，是临海市往东方向进出城区的过渡区域，交通十分便利。项目周边地块规划以住宅为主，并配套有中小学，目前周边商业、交通设施较为匮乏。

项目背景

临海市，是浙江东南沿海的一座古城新市，属沿海经济开发区。历史悠久，人文荟萃，经济、文化发达，至今仍保留着许多古遗址、古建筑和大批珍贵文物，1994年被国务院公布为国家历史文化名城。临海素有“小邹鲁”和“文化之邦”的美誉。此外，随着经济文化的发展以及徽商对当地的影响，徽州文化也走进了临海，对临海的建筑文化产生了影响。

项目定位

项目规划有低层合院、排屋、高层住宅、会所以及配套设施等。由于基地所处的特殊的地理位置，项目基本定位为“文化、生态、健康、和谐的现代宜居社区”，以提升和展示临海市的居住品质和面貌。

Location Overview

The Project locates on the Shuanglv Road northwest plot, in the north area of Linhai midland. The site lies against mountain in the northwest and faces villages in the southwest. The plot occupies northeast of Linhai, which is a transition region for going east to downtown, with convenient transportation. The peripheral planning plots are for residence with supporting facilities of primary schools and middle schools, however, it is short of commercial and public transportation supporting for the moment.

Project Context

Linhai is an ancient yet booming city close to southeast coastal area in Zhejiang, and it belongs to coastal economic open areas. It is a time-renowned city with flourish culture and economy. Many historic ruins, old buildings, ancient mausoleums and treasured relics are conserved here. Linhai was designated as National Historic and Cultural City by the state council in 1994. Linhai is also called “Small Zou Lu” and “Culture of the State”. In addition, with the development of economy and culture, Huizhou Merchants culture has entered and stationed in Linhai and has exerted great influence on local architectural culture.

Positioning

The project includes low-rise enclosure courtyard, townhouse, high-rise residence, club and supporting facilities. According to its geographic location, the project aims to build a modern livable community oriented by culture, ecology, health and harmony so that Linhai residential quality and city appearance will get a great improvement.

规划设计构思：古典“十”字形

结合山体与南北向水渠的自然现状条件，引入唐长安城里坊制城市规划概念，规划“十”字形的两条结构，其中山体位于西北区，高层位于东北区，南侧两区为低层住宅，形成南低北高的城市空间格局，不再是站在社区观自然山水，而是社区生长于自然山水之中。

总体规划设计：“十”字形结构

规划结合山体“十”字形结构将社区划分为除山体外的三个片区，三个区自成体系。南北景观轴是指利用现有南北向水渠为依托，在其基础上进行景观设计，形成南北滨水连续的景观轴，结合山体形成自然山水空间，将大自然引进社区。东西结构轴是以入口景观道路为基础，将地块南北划分为高层与低层的南北两区，结构轴与景观轴呈“十”字结构。

在总体视觉形态上南侧由方格网络组织各个里坊，北侧高层区点板结合，在东西向形成视觉通廊，将西侧的山与东侧的水结合，让高层地块真正生长于山与水之间，整个社区移步易景，美不胜收，中心景观结合“十”字轴巧插妙连，使整个社区充满了空间变幻。点式建筑结合滨水景观带布置，有效地形成景观渗透。

Planning Design: Classic Cruciform Layout

In the light of natural conditions of mountain and north-south water channel, the planning brings Chang'an alley system planning concept in the Tang dynasty and forms a cruciform structure. The mountain is in the northwest, the high-rise residence is in the northeast, and low-rise residences sits in the south two sections, constituting a spatial layout descending from north to south. The planning makes the community look like growing in natural landscape instead of building in it.

Comprehensive Planning Design: Cruciform Structure

The overall community is divided by cruciform mountain into three independent sections. The north-south landscape axis relies on water channel with resemble direction. Basing on the water channel, a successive waterfront landscape axis combining natural landscape is brought in the community. The east-west axis starts from entrance scenic road and divides the plot into high-rise section at north and low-rise section at south . The east-west axis and landscape axis intersect into a cruciform structure.

As for the overall visual modality, the southern sections consist of network alley system while the northern sections are composed of connections of points and plates. Along east-west visual vestibule, a combination of the mountain at west and water at east makes the high-rise residence grow in landscapes. The whole community is beautiful with changing scenery with moving steps, and the central landscape integrates well with the cruciform axis, producing a community with alternating spaces. Dotted buildings interact with waterfront landscape axis, bringing a landscape infiltration effect.

立面与造型设计：简化版古典徽派建筑

整个社区建筑立面和造型主要从人性化的角度出发，用建筑的语言拉近人和建筑之间的距离。丰富的建筑立面使建筑整体更富灵性，在深层次中挖掘建筑自古以来要求的人性化，提升了住宅作为居住空间的舒适性与变化性，演绎诗意栖居的高尚生活。

低层住宅采用简化后的中国古典徽派建筑的风格，建筑采用坡顶的形式，以增强、丰富建筑的第五立面，同时很好地解决保温隔热、防水的问题。建筑形象给人以舒展、稳重之感，水平线条的处理统一而不失丰富。作为对比，高层建筑采用现代中式风格。

低层住宅建筑外饰面材料以浅色涂料和深色青砖搭配，配以细巧、精致的线脚、暖色的花岗岩基座、浅色隔热玻璃，使不同的色彩和材料通过建筑体量竖向的变化相互穿插组合，在立面上形成层次感。

户型设计

低层住宅大部分采用围合院落的形式布置，高层住宅则主要采用一梯两户、局部一梯三户和一梯四户的单元形式。

设计以人为尺度，对建筑平面进行立体思考，将家庭结构和生活习惯奉为设计标准，使人在同样的单位面积里能享受到更合理的布局，更自由的生活。平面套型设计根据总体布局灵活多变，保证主要的生活用房，包括厨房、至少一个卫生间都有直接自然通风和直接自然采光。平面分区明确，流线顺畅，尽量避免住户间的视觉与声音的干扰。同时所有公寓的公共空间，如大堂、电梯厅、楼梯间等，都做了全明设计，以体现小区的人性化。

Facade Design: Minimal Classic Hui-style Architecture

The overall facade and modeling design are from human-orient perspective to shorten the distance between people and building by architectural language. Diversified facades stimulate architectural spirituality and deeper humanization. The design brings comfortable and diverse living spaces displays a poetic noble life.

The low-rise residence facade adopts minimal classic Hui-style, such as sloping roof which enriches the fifth elevation and solves heat, insulation and water-proof issues. The parallel lines are unified yet not monotonous. The residence leaves a stretching and calm image. By contrast, the high-rise residence facade employs modern Chinese style.

The low-rise facade is painted with light-color coating matching dark brick, delicate skintle, warm-color granite plinth and light-color insulated glass. Various colors and materials interact vertically, creating rich layering.

House Type

The low-rise residences are in enclosure courtyard layout, while the high-rise residences are in unit forms of two, three or four families sharing one floor.

The design respects family structure and living customs as standards to arrange proper layout so as to make people enjoy free life in the same unit area. House type sizes are various according to general layout, which guarantees main rooms like kitchen room and at least a toilet to get natural ventilation and lighting. The plane division is clear with smooth circulation while prevents visual and noise interference. Every public space, for instance lobby, elevator room, stairs, all equips lighting facilities, so this is a real human-oriented community.

小贴士

马头墙与徽州壁画是徽派建筑的重要组成部分。马头墙又称风火墙、防火墙、封火墙，特指高于两山墙屋面的墙垣，是山墙的墙顶部分，形似马头。其构造随屋面坡度层层迭落，呈阶梯状，墙顶挑三线排檐砖，常见的马头墙有一叠式、两叠式、三叠式、四叠式，砖墙墙面以白灰粉刷，墙头覆以青瓦两坡墙檐，白墙青瓦，明朗而素雅。而徽州壁画则描绘在徽州古民居的屋檐下和门楼、窗檐上下，以一个墙面多幅壁画为一体，以工笔写意为技法，以美化徽州民居外墙壁。

Tips

Dwarf wall and Huizhou mural are important parts in Hui-style architecture. Dwarf wall is also called firewall that overtops gable roof as the top of gable wall, and it is shaped like a horse' s head. The dwarf walls descend with slope roof in a terrace formation. One stile, two stiles, three stiles and four stiles are commonly seen in Hui-style dwellings. White wall matches black eaves and tiles depicting a bright and elegant painting. Huizhou murals are decorated under eaves, on gatehouse, around sunblind. Several murals are painted on a wall as a whole, and its meticulous techniques beautify out walls of Hui-style dwellings.

移植徽派建筑元素 现代与传统碰撞

海口中华坊

Hui-style Elements Transplant Modern and Traditional Collision

Zhong Hua Fang, Haikou

开发商：海南金源世纪房地产开发有限公司 建筑设计：海南华磊建筑设计咨询有限公司 项目地址：海口市南海大道11-1号
占地面积：200 000平方米 建筑面积：290 000平方米 容积率：1.2 绿化率：40.02% 采编：谭杰

Developer: Jinyuan Century Real Estate Development Co., Ltd., Hainan Architecture Design: Hualei Architecture Design Consulting Co., Ltd., Hainan
Location: No. 11-1 Hainan Avenue, Haikou Site Area: 200,000 m² Building Area: 290,000 m²
Plot Ratio: 1.2 Greening Rate: 40.02% Contributing Coordinator: Tan Jie

海口中华坊虽地处海南，却将江南建筑中的婉约、素雅表现得淋漓尽致。项目设计移植徽派建筑中的马头墙、坡屋顶等建筑元素，以全新的方式演绎徽派建筑，将传统中式建筑语素与现代技术相结合，在传承街坊与院落文化的同时，使灰白色墙体、片状的构件、通透的大玻璃搭配钢架和玻璃组成的塔窗与坡屋顶进行巧妙结合，形成现代与传统的碰撞。

Although the project locates in Hainan, it vividly expresses Jiangnan architectural graceful and elegant features. Wharf wall and sloping roof are Hui-style architectural elements that are transplanted in the project to display Hainan Hui-style architecture. The design combines traditional Chinese style architectural language and modern technologies to pay respects to lane system and courtyard culture. Hoary wall, flake-like component, large-sized transparent glass on tower windows smartly echo the sloping roof. The project is a masterpiece of modern and traditional collision.

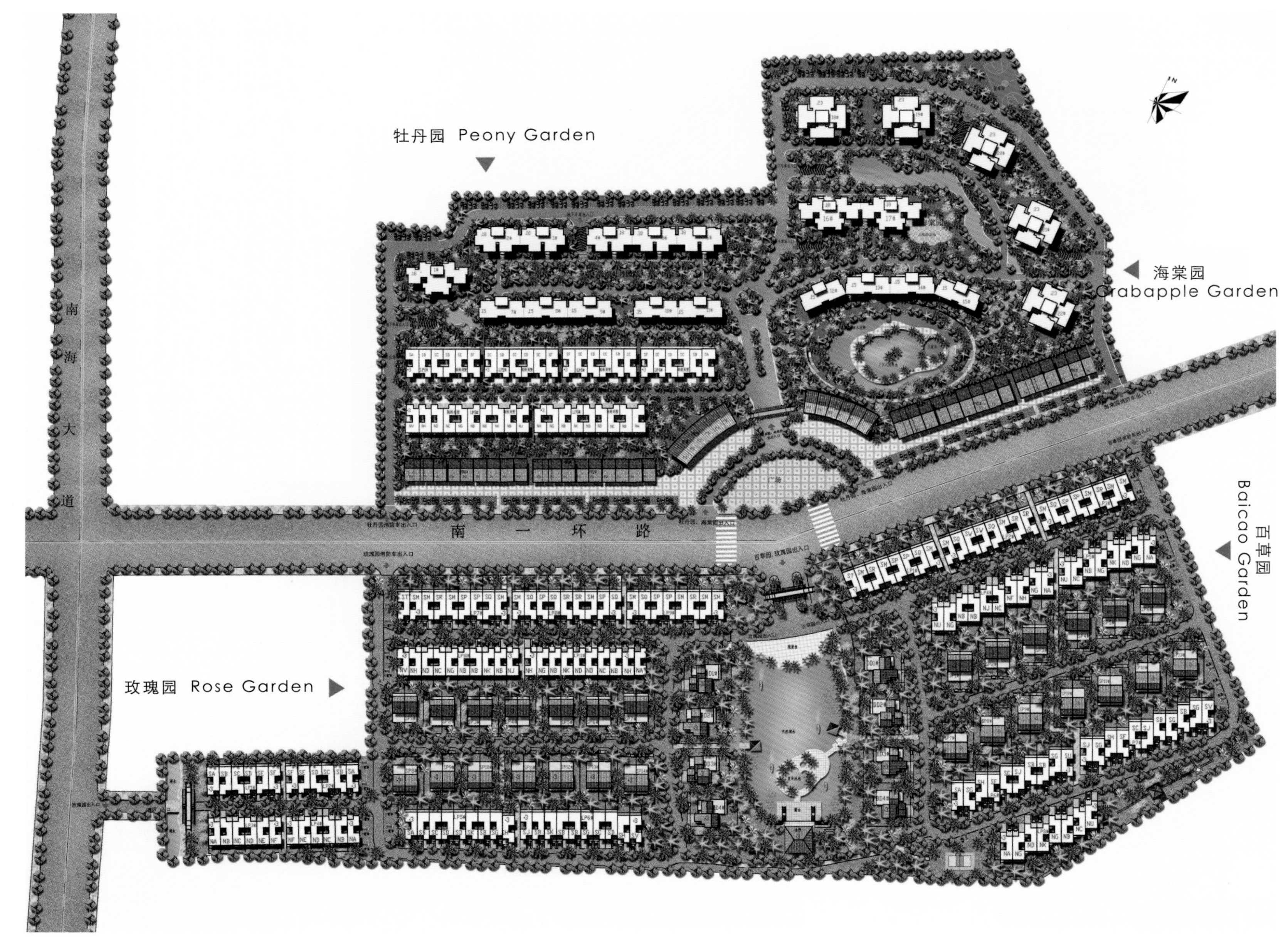

总平面图 Master Plan

区位分析

项目用地位于海南省老城开发区快速干道23公里处，北边为南一环路，西边为南海大道，用地两面环市政规划路。项目所处地段交通便利，地势由南北面逐渐向东南面抬高。玫瑰园及百草园的社区中央有一个约7 000平方米的天然湖。

项目背景

海口是个容纳百川的城市，对于外来文化的吸收和包容性很强。而庭院是中国民居的灵魂，在中国人的传统中更像是一个大的开放的起居厅。西方住宅以起居室为核心展开布置，中国传统居民则以院落为中心。

Location Overview

The project is 23 km away from Expressway of the old town development zone, Hainan. South First Ring Road lies in the north; Hainan Avenue is in the west; and the other two sides are municipal planning roads. The project site enjoys convenient transportation conditions. Its terrain is gently ascending from northwest to southeast. In the center of Rose Garden and Baicao Garden communities lies a natural lake about 7,000 m².

Project Context

Haikou is an inclusive city for external and diversified cultures. Courtyard is a spirit in Chinese dwellings, and it is like a large and open living hall in the mind of tradition. While western residence takes living room as focus, the Chinese dwelling regards courtyard as the center.

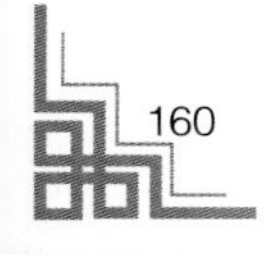

玫瑰园，百草园竖向设计图

Rose Garden, Baicao Garden Verticel Plan

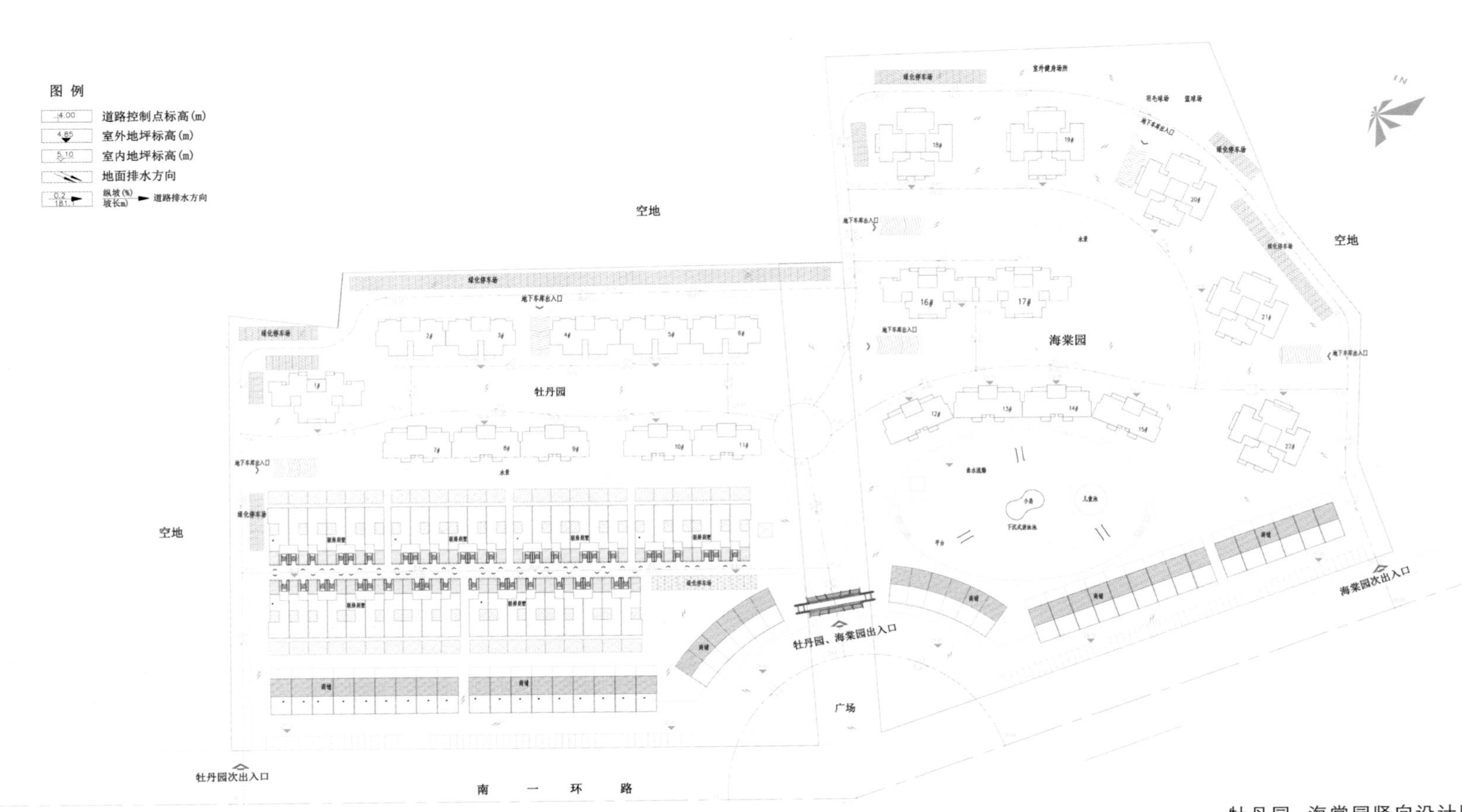

牡丹园、海棠园竖向设计图

Peony Garden, Crabapple Garden Verticel Plan

设计思想

项目定位为现代中式住宅，意图通过对中国传统民居建筑的继承与创新，对本土文化的关心与现代生活的契合点，创新适合中国人居住的建筑形式。既萃取传统文化精髓，又体现时代建筑精神，成为可以传世的经典。

对“街坊”的继承：栋栋相连的门楼，亲切的邻里、街坊。

对“院落”的继承：民居的前庭院是入户的前奏空间，使建筑有了一个由室外公共空间进入室内空间的过渡、承接与缓冲。设计利用两排别墅间的日照距离形成的后私家花园，使私密空间得到了进一步的延伸与扩展。丰富的院落空间补充了室内的不足，成了人们在家中走进自然、享受阳光的最佳场所。

对中国传统民居建筑造型及色彩元素的继承：在项目的创造中，设计力图通过采用青、灰等建筑中常见的青色淡雅色彩创造一种舒适、休闲、放松、平和的家居环境，引入在南方居民建筑中较常用的大面积留白墙面，约1/3白墙和2/3灰墙的搭配在建筑立面上形成了较为强烈的对比，改变了传统民居给人严肃、压抑的感觉，使其活泼大方，符合现代人的审美观，并运用现代审美观进行了艺术处理、简化、变形，在细部处理上进行了推敲，使其符合现代人的审美观及现代施工工艺。

总体布局

项目利用小区的天然湖成为中心轴线景观，以大面积的湖体为主形成整个小区的核心区域。泳池在概念上设计成为湖体的一部分，会所“漂浮”于水面上，使湖体、泳池与会所连成一体，成为张弛有度而又怡人的社区水公园。独栋别墅沿着湖面周边布置，联排及双拼平行红线布置，利用两排别墅的间距形成街坊，给居住者提供了一个交往的空间，使邻里之间具有较强的亲和力。

Design Conception

Zhong Hua Fang is a Chinese style residence. The project inherits and innovates Chinese traditional dwellings, and its design concerns local culture and modern life, creating a proper architectural form suitable for Chinese people. It is a classic project containing traditional cultural essence and modern architectural spirits, which can be handed down from age to age.

Inheritance for Lane System: Detached houses are standing side by side, forming intimate neighborhood.

Inheritance for Courtyard Framework: The front court is a prelude space for residents entering home, and it is also a transition or a buffer for people going in their private space from external public area. The rectangular shadow between two rows of villas forms a rear private garden, which extends private spaces and complements limited interior spaces. People can enjoy nature and sunshine at home.

Inheritance for Chinese Traditional Dwelling modeling and Color: The design takes quiet and elegant blue and gray to create a comfortable, relaxed and peaceful household environment. In southern China, dwelling house has large-sized blank metope, and the project uses one third white wall and two thirds gray wall to alleviate the solemn and depression feelings in traditional dwellings. The adjustment motivates the atmosphere and conforms to modern aesthetics. The design utilizes modern aesthetic art to simplify and transform details, boasting modern construction technologies.

Overall Layout

The natural lake forms a central axis landscape, and it is also the core area of the community. Swimming pool is a part of the natural lake, and a community club “floats” on the lake. The layout makes the community to be a water park. Detached villas sit around the lake, townhouses and semi-detached villas are arranged on property line. The distance between two rows of villas forms a lane where residents can communicate freely.

建筑造型

设计借用灰白色墙体、马头墙、台阶、宅门、屋檐线条、窗口雕花等传统元素，使简单平面构成的空间形态变得丰富和富于美感，形成户户相对的街道式排屋格局。在别墅主体的屋顶形态上，以两坡屋顶组合为主，构成坡顶与坡顶之间的高低错落。局部通过木构造的结构勾勒出中国传统屋顶的形态，形成青瓦面与镂空屋架的虚实对比，屋面出檐的大小、行制以及檐下空间的变化，都统一在具有相同坡度的街区屋面形态中。而在入口门头的屋面形态处理上，屋脊采用的装饰是一种介于南方精巧秀丽的翘屋脊与北方平直朴实的平屋脊之间的形态。每一处入口空间都在立面上做了精细的处理，形成相对独立的构图单元。

Architecture Modeling

Hoary white wall, wharf wall, entrance stairs, house door, roof line, carved window are traditional elements that enrich and beauty spatial forms. Houses are face to face, forming a lane-like townhouse layout. Villas cover duo-pitched wooden roof in different heights to form dislocated Chinese traditional roof configuration. Gray tile and hollow roof truss constitute virtual-real comparison. Eave hood size and the space under the eave are in similar gradient with the whole house roofs. As for the door head roof, it coordinates southern exquisite tilted ridge with northern flat roof features. Every entrance space has unique facade design, and all of them are independent photographic units.

单体设计

在独栋别墅中，起居室做成了两层通高，提高了居住品质，创造了较为丰富的室内空间；在双拼和联排别墅中，起居室地面下降1.2米，使其和其他生活场所之间有了一个空间上的划分，同时使起居室空间变得更大、更宽敞。此外，还利用每户院子内外的高差，布置地下车库，汽车从街道入库后，业主可通过防火门直接进入室内，满足了现代人的生活习惯及心理要求。以前庭，中庭，后院的中式建筑布局，形成四水归堂，藏风聚气，旺人兴旺的吉祥寓意。

前庭既是入户的空间过渡、又是住户心理的调节和缓冲，而且在功能上把工人房的入户放在前庭，和住户不发生干扰，最大化地提供住户私密性的要求。天井成为藏精聚气、通天透地的小型围合空间，除保证使用房间的采光通风，对住户还考虑了积极和动态的生态性能。后院更是住户闲情逸致之下陶冶情操、修生养性的绝佳场所。作为居者，沿着门前的台阶而上，进入别墅大门，这里有一个约20平方米的前庭，为了增加私密性，后花园有围墙将其环绕，并可以看到别墅首层里布置的12平方米左右的内天井。这些细部的设计能让人们在体验住宅的现代感、舒适感的同时，觉察到一丝的古意，感受到一份细腻而怡静、充满中国文化和哲学的生活。

交通组织

中华坊在设计中采用了紧凑的前后排共用一条街道的街坊式布局，并用自然的曲线道路加以组织，创造安逸静谧的居住空间。小区中间横贯南北保留天然人工湖及湖边步行景观带，为居民提供了一个休闲、活动场所。

Monomer Design

In the detached villas, living room are two floor high to provide a potential of a rich interior space; In the semi-detached villas and townhouses, living rooms sunken 1.2 m to differ from the rest spaces and enjoy a spacious space. Basement garage utilizes the height discrepancy of inside and outside yards, so residents can enter in from basement fire door after parking their car. The pedestrian route meets modern people's living habits. In addition, the Chinese layout of front court, medium court and rear yard implicates auspicious meanings.

The front court is a transition space for residents adjusting moods. Maid's room is settled in the front court to prevent interruption and maximize private spaces. Patio is a tiny vertical enclosure space to bring lighting and ventilation and offer active ecological effects. The rear yard is for relaxation, recreation and amusement. Stamping on entrance stairs and through the gate appears the front court about 20 m². In order to protect privacy, the rear garden sets a fence, and the patio about 12 m² is embraced by the fence. These design details express modern comfortable habitation requirements. Living here, people can taste an ancient conception that it is an exquisite and tranquil life with cultural and philosophic meanings.

Communication Arrangement

The project takes compact lane system layout that two rows of villas share the same lane, and they are connected by meandering roads to offer quiet habitation spaces. The natural lake lies from south to north in the community, and the lakefront pedestrian landscape belt is a good place for people resting and playing.

海南徽派意蕴 禅意度假居所

海南远洋华墅一期

Hainan Hui-style Expression Zen Holiday Residence

Phase I of Sino Ocean Zen House, Hainan

开发商：远洋地产（海南）有限公司 建筑设计：美国Robert Hidey Architects 建筑设计事务所/日兴设计•上海兴田建筑工程 景观设计：新加坡贝尔高林私人有限公司
项目地址：海南省海口市西海岸盈滨半岛 占地面积：158 746平方米 建筑面积：109 000平方米 容积率：0.33 绿化率：47% 采编：谭杰

Developer: Sino-Ocean Land Hainan Branch Architecture Design: Robert Hidey Architects, Architects Nikko
Landscape Design: Belt Collins International (Singapore) Pte. Ltd. Location: Yingbin Peninsula, West Shore, Haikou, Hainan
Site Area: 158,746 m² Building Area: 109,000 m² Plot Ratio: 0.33 Greening Rate: 47% Contributing Coordinator: Tan Jie

"蓝天白云、水清沙白"是人们向往的度假之地和生活之地，海南因海岸和气候等多个因素得到了旅游度假者和居住者的青睐，海南的建筑多以度假和第二居所居多。海南远洋华墅将中式奢华而淡雅的徽派风格与海南度假风情相结合，以东方禅意庭院塑造充满诗情画意的园林，以中国古典音律"宫商角徵羽"和古琴名曲作为景观的主题，在海口打造出蕴含徽派建筑文化内涵和气质的休闲、度假居所，与"阳光海南、度假天堂"相呼应。

Blue sky, white cloud, clear water and white sand form a desirable holiday and residing resort. Due to pleasant climate and beautiful coast, Hainan gets favor from holiday tourists and inhabitants, hence, most of buildings in Hainan are for holiday and a second residence. The Sino Ocean Zen House integrates sumptuous and elegant Hui-style and Hainan holiday flavor. It is a oriental courtyard full of poetic sentiment. Traditional ancient five tones and ancient music masterpieces are taken as landscape themes to build a Hui-style cultural holiday and recreational residence, echoing the comment that "the sunshine Hainan is a holiday heaven".

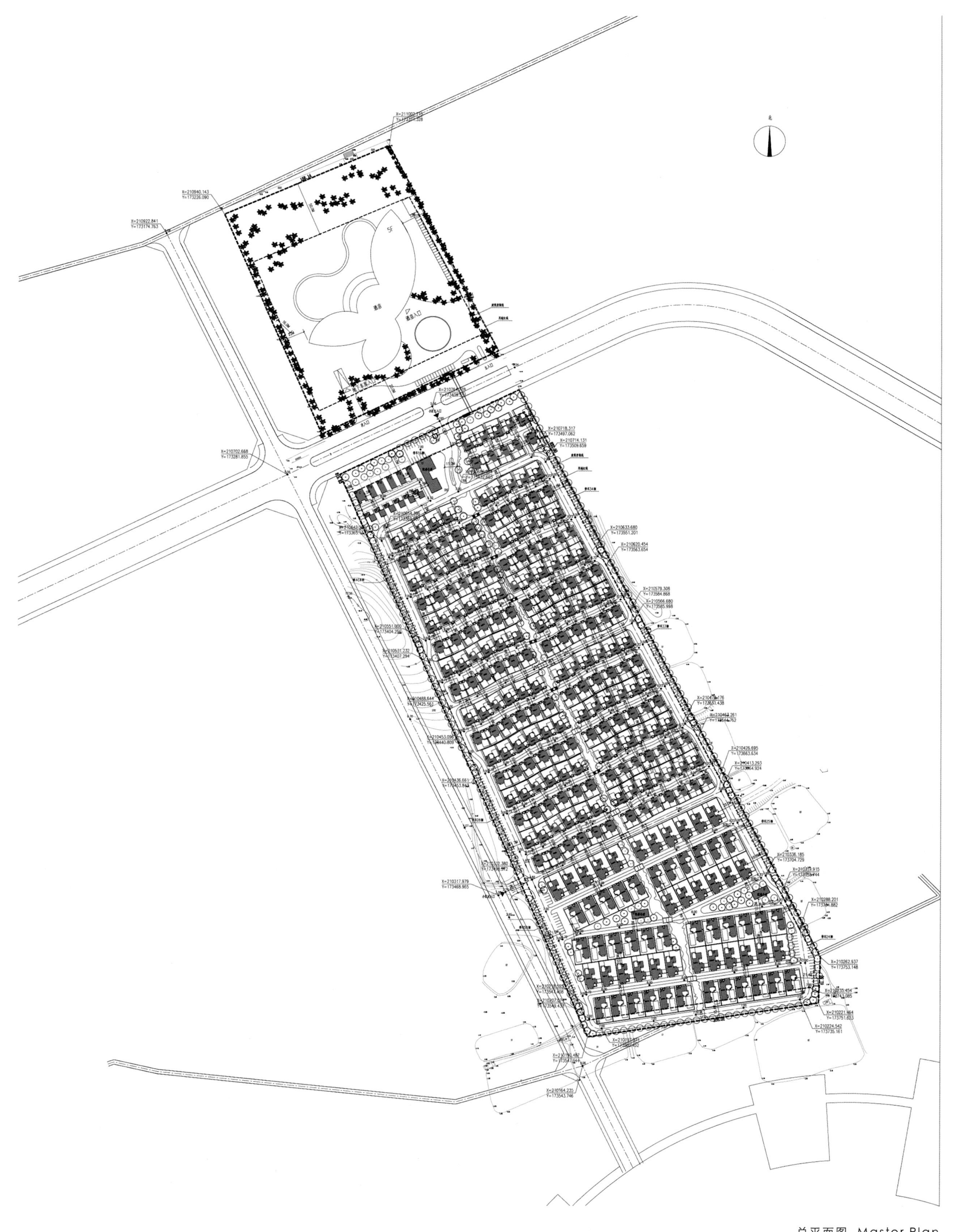

总平面图 Master Plan

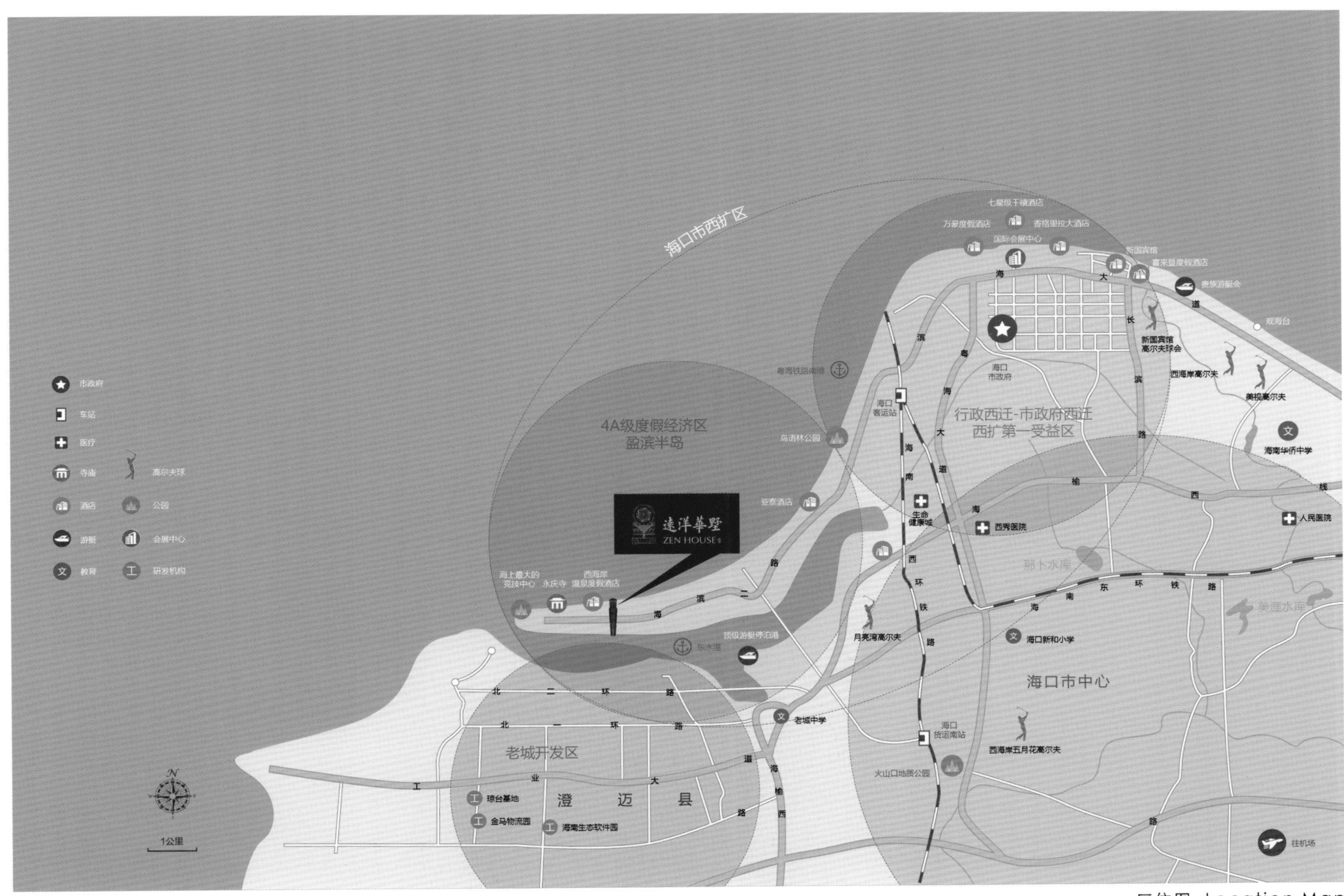

区位图 Location Map

区位分析

项目位于海口市西海岸盈滨半岛旅游度假区，北临琼州海峡，南靠内海。周边配套完善，交通便利。

项目背景

徽派建筑注重采光，以木梁承重，以砖、石、土砌护墙，以堂屋为中心，广泛采用砖雕、木雕、石雕，给人奢华而淡雅之感。而其园林、庭院的设计，又有中国泼墨山水画之感，赢得了居住者的青睐。

院落是传统建筑的灵魂，东方禅意庭院起源于中国，最具代表是中式泼墨山水式的庭院和日式洗练素描式的庭院。庭院设计秉承中式传统哲学和绘画艺术，重视诗画情意，于情于景，创造至美意境。

设计构思

项目设计初衷是将徽派的村落及传统的文化元素融入海口的地域和环境之中，打造徽派特色和休闲、度假相结合的第二居所住区。设计将建筑、院落、景观在一个宜人的居住尺度中有层次地铺陈展开，体现出中国传统园林"小中见大"的特点。

Location Overview

The project locates in the Yingbin Peninsula Tourist Resort, west shore of Haikou. It faces Qiongzhou Strait in the north and backs against the South China Sea with complete peripheral supporting facilities and transportation conditions.

Project Context

Hui-style architecture highlights lighting and features in wooden pillar as load bearing column, parapet wall built by brick, stone and clay; central living room as focus, and brick carving, wood carving and stone carving used widely. These elements bring sumptuous and refreshing flair. The garden and courtyard are like an inkwash drawing scroll, making dwellers can't help being attracted by it.

Courtyard is the soul of traditional architecture. Oriental zen courtyard originated from China, and Chinese inkwash-esque courtyard and Janpanese sketch-esque courtyard are representatives. The Sino Ocean Zen House obeys Chinese philosophy and painting art to create a space full of poetic sentiments.

Design Conception

The design aims to integrate Hui-style village and traditional cultural elements in Haikou to build a recreational and holiday residence in Hui-style. Buildings, courtyards and landscapes are organized in a proper scale and present in a reasonable hierarchy, incarnating a traditional Chinese garden feature that you can see a world in a spot.

规划设计：以“和”理念布局

项目基地地形形似古琴，纵贯盈滨半岛南北两极。北区直面大海，规划养生会馆、滨海娱乐体验中心，尽揽一线双湾海景，可独享23公里黄金海岸，与南区内海港湾遥相呼应；南区则以“和”的理念布局240席独栋别墅，规划蕴涵风水学术，筑园契合古琴五音韵律，通过绿绮桥、号钟桥、绕梁桥、独幽桥、焦尾桥、中和桥、飞泉桥联通“宫商角徵羽”社区组团，充分演绎邻里友好空间。

建筑立面设计：徽派风格

建筑以黄土作墙，采用木屋黑瓦设计，依山傍水、环境宜人，以徽派建筑风格营造小桥流水人家的意境，形成建筑、人与自然的和谐统一，营造回归心灵、亲近自然的氛围。

建筑单体设计：前二后三

独栋的设计采用“前二后三”的竖向设计，相对入户道路层数是两层，地下室向下沉院落面打开，相对下沉建筑层数为三层。设计采用8米台地二层设计，灵活定制5米挑高空间，实现360度全景采光，以三进制中式待客礼仪，结合超大花园露台和庭院温泉入户，传承中式文化构思，将中式建筑精髓凝练在每一处角落。

Planning Design: Harmony Concept

The site is like a Chinese seven-stringed plucked instrument lying across from north to south. The northern part facing the sea builds a Health Maintenance Club and a Coastal Entertainment Experience Center. They enjoy double-bay seascape and 23 km golden coastline, echoing the southern part in a distance. The southern part tiles 240 detached villas and their layout complies with geomantic omen and ancient rhythms. Chinese ancient five tone named community clusters “do re mi sol la” are connected by Lvqi Bridge, Haozhong Bridge, Raoliang Bridge, Duyou Bridge, Jiaowei Bridg, Zhonghe Bridge and Feiquan Bridge, creating a harmonious neighborhood space.

Facade Design: Hui-style

Wooden houses with yellow clay wall and black tile stand by waters and mountains, enjoying pleasant environment. Bridge, water and dwelling form Hui-style feelings. Architecture, human being and nature are organized in a harmonious unity, bringing a natural atmosphere to help people return to the heart.

Architectural Design: 2+3 Monomer Layout

The monomer design adopts 2+3 vertical layout. It is two floors if looks from the entrance, while it is three floors because the basement inlays at the floor of sunken courtyard. The 8 m high house is divided into two floors with 5 m high raised ceiling space to allow 360º panoramic daylighting. There is a living room with three doors, a super-sized terrace and a hot spring in the courtyard. The design inherits Chinese cultural conception and extracts Chinese essence presenting at every corner.

景观设计：以音律为题

景观设计采用新中式园林设计方式，崇尚建筑、山水、植物的融合，“落轿入府”的南区入口广场、街巷式过渡庭院、半私密的街巷院落体系、入户迎客庭院、围合式私密主庭院的五重庭院设计，采用街巷式布局、院落式空间，讲究布局的层层递进，强调空间的延展性。同时，采用全冠移植技术，不远万里从世界各地遴选苏铁、文殊兰、刚竹、黄槿等200余种珍贵名木，营造错落有致的空间质感。

中央景观以水为设计主题，水作为小区的景观基本元素，以水形成住区景观，以传统音律的“宫商角徵羽”和古琴名曲作为景观的主题，串连起整个小区，自然、生态的景观主轴仿佛是一曲悠扬的古曲在演奏。鱼骨状的规划结构简洁明确，中央的主轴串联起一条条的街巷，街巷内回到了传统村落的宅院模式，门头、院墙、照壁让人宛如置身徽派的古村落之中，组团内的独栋通过院落划分将场地完全分配到每家，延续中式园林幽深细腻的情致。

Landscape Design: Theme of Melodious Tone

Landscape design applies neo-Chinese garden style, and emphasizes the combination of architecture, landscape and plant. South Entrance Plaza, Lane Transition Courtyard, Semi-private Lane Courtyard, Home-entry Courtyard and Enclosure Private Main Courtyard form quintet courtyards. The lane system layout and courtyard space are in a progressive layer form, highlighting an extending space. In addition, in order to create a dislocated space, the landscape design introduces crown transplantation technology, and chooses more than 200 species of precious rare plants from the world, such as sago cycas, poison bulb, phyllostachys viridis and hibiscus tiliaceus.

Central landscape takes water as theme and basic landscape element. Traditional ancient five tones and ancient music masterpieces are taken as landscape themes to series up the whole community. The natural and ecological landscape axis is like a melodious ancient melody. Herringbone layout is concise and explicit, and the main axis connects strips of lanes. The dwellings in the lanes appear traditional village-like courtyard mode. Door head, courtyard wall, screen wall bring people to an old Hui-style village circumstance. Every monomer shares courtyard space in clusters, extending a Chinese garden spirit.

京鲁风韵

The Charm of Beijing and Shandong

京鲁包括北京、河北、山东等地，这些地方都有着丰厚的历史底蕴，尤其是北京，它是有着3 000多年历史的国家历史文化名城，在历史上曾为六朝都城，建造了许多宏伟壮丽的宫廷建筑，使北京成为中国拥有帝王宫殿、园林、庙坛和陵墓数量最多以及内容最丰富的城市。北京四合院是老北京最主要的民居建筑，源于元代院落式民居，以正房、倒座房、东西厢房围绕中间庭院形成平面布局的北方传统住宅的统称。随着经济文化的发展，北京四合院对周边乃至全国的建筑文化都有着深远的影响。山东是儒家文化发源地，儒家思想的创立人孔子、孟子，以及墨家思想的创始人墨子、军事家吴起等，均出生于鲁国。

在此我们从历史文化、现代科技等角度来分析当地建筑的地域特色及在经济文化的发展影响下的演变。如尊重历史文脉的北京钓鱼台艺术酒店，设计延续国子监历史与文化脉络，曲阜香格里拉大酒店则是从山东的儒家文化中寻找设计灵感，演绎出全新的具有儒家文化的建筑。这些建筑还原当地的历史文化，同时又结合了当下的材料与设计进行了全新的诠释。再如泰禾北京院子的设计采用了传统的坊巷规划设计，同时也采用了新的材料和技术，不仅传承老北京四合院文化的传统情怀，同时也营造出了富有现代气息的生活空间。

This chapter includes architecture works in Beijing, Hebei and Shandong that all have rich historic foundations. Beijing in special, is a historic and cultural city owning more than 3,000 years history. It had been the capital for six dynasties, so many magnificent palaces were built there, which has made Beijing enjoy the most imperial palaces and gardens, temples, altars and mausolea in China. Beijing quadrangle courtyard is the most important residential architecture in Beijing, the dwelling was derived from courtyard dwelling at Yuan dynasty. The quadrangle courtyard is a general term for a northern traditional residence consisting of main room, the rooms opposite the main room, wings at both sides of the main room and a central yard in plane layout. With the development of economy and culture, Beijing quadrangle courtyard influences peripheral and even all the architectural culture in China. Shandong is the birthplace of Confucian culture. Many historic celebrities, for instance the Confucianism founder Confucius, Mencius, the Mohist thought founder Mo-tse and militarist Wu Qi, etc. born in the State of Lu.

Here we analyze local architectural characteristics and their evolution under development of economy and culture from the aspects of history, culture and modern technology. Beijing Diaoyutai Art Hotel respects historic texture, so the design extends Imperial College context; Qufu Shangri-La hotel design inspires from Confucian culture in Shandong, and it presents a brand-new Confucian cultural building. Utilizing architecture restores local history and culture, and meanwhile combines modern materials and design methods to interpret local historic and cultural connotations. Beijing Cathay Courtyard adopts traditional lane and alley system planning and modern advanced materials and technologies to inherit Beijing quadrangle courtyard culture and create a modern living space in the mean time.

延续国子监文脉 传达新中式情怀

北京钓鱼台艺术酒店

Extend Imperial College Context　Convey Neo-Chinese Feelings

Diaoyutai Art Hotel, Beijing

物业公司：钓鱼台美高梅(北京)酒店管理有限公司　建筑/景观设计：美国波士顿国际设计集团(BIDG)　项目地址：北京市东城区国子监文化保护区
占地面积：16 352平方米　建筑面积：40 214平方米　容积率：1.56　绿化率：27.34%　采编：康欣

Management Company: Diaoyutai MGM Hospitality　Architecture/Landscape Design: Boston International Design Group (BIDG)
Location: Conservation Zone of Imperial College History & Culture, Dongcheng District, Beijing　Site Area: 16,352 m²　Building Area: 40,214 m²
Plot Ratio: 1.56　Greening Rate: 27.34%　Contributing Coordinator: Kang Xin

国子监街区作为北京重要的历史风貌保护区，其街道特色和建筑风貌独具历史韵味，而重要历史建筑的大轴线、小组团布局以及片区的传统胡同及四合院建筑肌理，是国子监街区建筑的突出特征。鉴于所在区域的历史文化底蕴，项目设计以精心设计的中轴线串联整个建筑群体，并尽力保留北京传统胡同与四合院的建筑肌理，在材料、形态、风格等方面契合周边区域环境，同时运用简约手法，对中式风格进行了现代演绎，从建筑、景观到室内设计无不延续北京地区的历史文脉，形成古典与现代的碰撞。

The Imperial College Street is an important historic conservation zone in Beijing with unique historic flavor. The great axes of significant architecture, small cluster layout and architectural texture of traditional Hutong and quadrangle courtyard are prominent features in the street. Basing on profound historic and cultural foundations, the Diaoyutai Art Hotel sets a central axis to link up the total architecture groups meanwhile retains the texture of Beijing Hutong and quadrangle courtyard. It employs materials, colors, shapes and styles that match those of the surroundings. The renovation is a modern expression for Chinese style, so the architecture, landscape and interior designs are defined with modern Chinese style, perfectly presenting a collision between classics and modernity.

景观设计

景观设计处处体现中国园林设计的精华思想，营造“移步换景”的效果。设计突出“水”主题，把多种形式的水景精妙地组合在一起，形成“北国水园”。主入口空间采用“重门”的形式，体现从动态的、喧嚣的外部都市到静态的、幽静的内部酒店的空间过渡，另外还通过增加双层墙面、乔木、水景，令整个视野在远、中、近景的分布上层次分明，具有鲜明的仪式感。

Landscape Design

In landscape design, the hotel fully embodies the essence of Chinese garden design, and creates an effect that scenery changes with every moving step. Water is highlighted in the design, and diverse waterscapes are masterly combined, forming a “northern water garden”. The main entrance flaunts layers of doors, creating a transition from the dynamic and bustling outside to the static and tranquil inside. In addition, double-layer wall, arbor and waterscape are used to create a clearly layered view ranging from distant, medium and close shots, forming a distinctive ritual sense.

中国经验的现代性创造
Modernity Creation of Chinese Experience
大都美术馆馆藏作品陈列展（二）
大都美术馆
Dadu Museum of Art

空间设计

客房设计延续国子监的历史与文化脉络，突出中式文化主题。设计对大堂空间进行了层高、开间、进深比例的重新塑造，通过最少的建筑修正，减轻了原先压抑笨重的感觉，取而代之的是高敞和有序。中式情结的元素巧妙嵌入空间的各个细部，使项目的文化品质得到更高的升华。珍品和宝书的引入，亦为项目增添了无与伦比的高端特色。

Space Design

As for guest room design, the hotel remains historical and cultural context of the Imperial College, highlighting Chinese culture. The floor height, width and depth of the lobby space are reshaped. The former repression and cumbersomeness are replaced by spaciousness and orderliness through minimum amendments. Chinese-style elements are skillfully inlaid into details in the space, enhancing cultural connotation in the hotel. Moreover, precious collection and treasured book add uncomparably upscale features to the hotel.

传承四合院形制 演绎新街巷院落文化

泰禾北京院子

Inherit Quadrangle Dwelling Layout　Present New Street Courtyard Culture

Cathay Courtyard, Beijing

开 发 商：北京中维泰禾房地产开发有限公司　项目地址：北京市朝阳区孙河乡北甸西村　建筑设计：北京翰时国际建筑设计咨询有限公司
用地面积：58 888平方米　建筑面积：104 985平方米　容积率：1.1　绿化率：30%　采编：张培华

Developer: Beijing Zhongwei Thaihot Real Estate Development Co., Ltd.　Location: Beidianxi Village, Sunhe Country, Chaoyang District, Beijing
Architecture Design: A&S International Design　Site Area: 58,888 m²　Building Area: 104,985 m²
Plot Ratio: 1.1　Greening Rate: 30%　Contributing Coordinator: Zhang Peihua

在传统中国建筑文化中，院落文化是不可缺少的建筑元素，北京院子传承北京四合院的“城市里坊”形制，创新采用“三街五巷八坊”的规划策略，将传统街巷与院落相融合，原汁原味地呈现了传统的坊巷规划理念。建筑设计虽采用新中式的风格，但木色凸花窗、中式屋檐、镶铜雕花木门及砖雕等传统经典装饰的点缀，在现代建筑的构造方式下，展现中式意境与现代时尚的结合，与北京厚重的中国文化底蕴和走在发展前沿的气质相呼应。

In traditional Chinese architectural culture, courtyard is an indispensible architectural symbol. The Cathay Courtyard inherits quadrangle dwelling “urban lanes and alleys” layout and innovates a “three streets, five lanes and eight alleys” plan. The design adopts neo-Chinese style and traditional ornaments such as carved wooden window, Chinese style eave, brass-insert carved wooden door and brick carving as decoration. Using modern techniques displays the combination of Chinese style and modern so as to echo Beijing profound cultural foundations and the cutting-edge developing trend.

总体规划平面图 Site Plan

A-K轴立面图 A-K Axis Elevation

K-A轴立面图 K-A Axis Elevation

1-1剖面图 1-1 Section

2-2剖面图 2-2 Section

3-3剖面图 3-3 Section

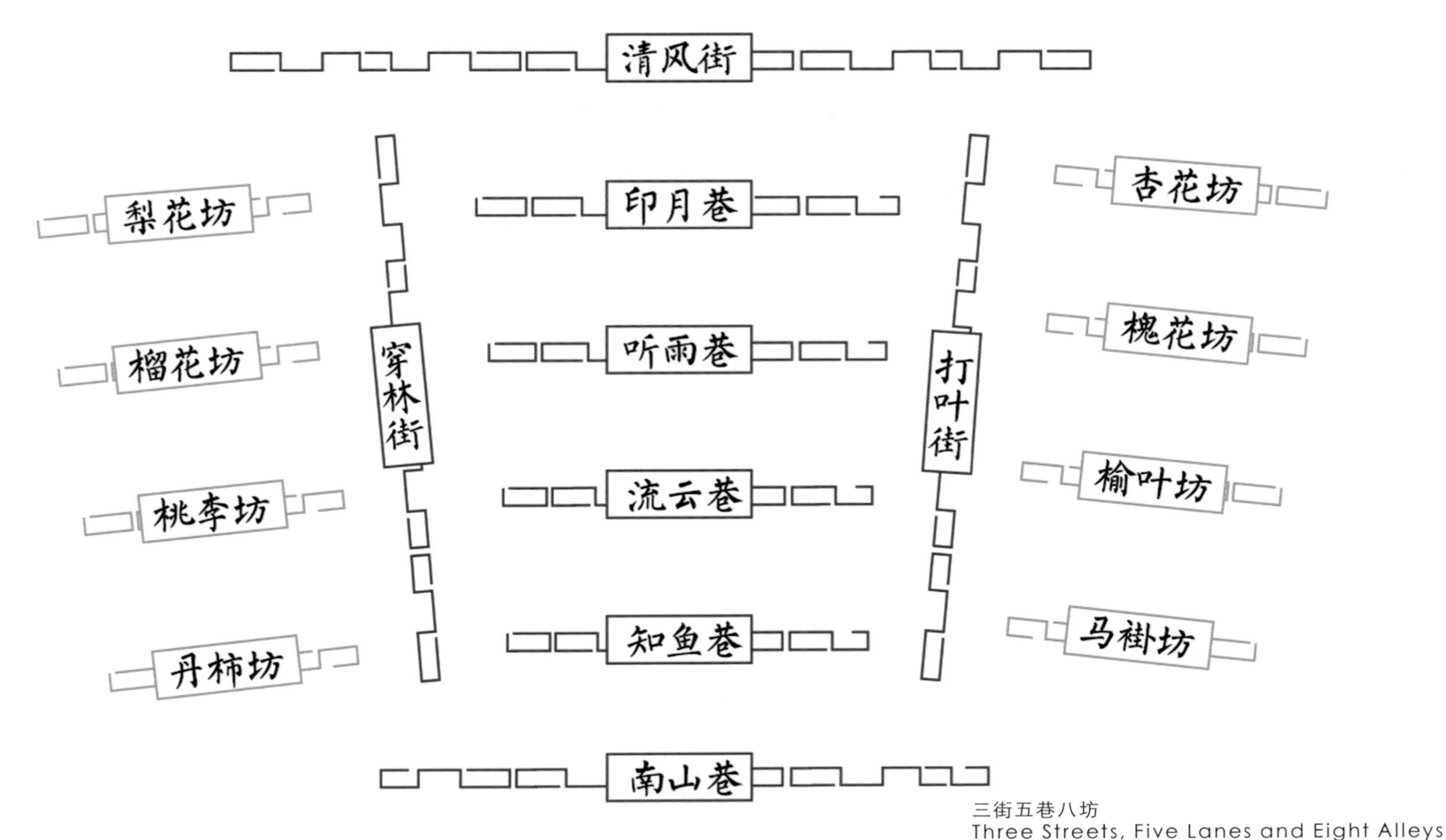

三街五巷八坊
Three Streets, Five Lanes and Eight Alleys

区位分析

项目紧邻三条核心高速路——京承、京平、机场高速，另有顺白路、顺黄路纵贯其中，紧邻地铁15号线，拥有立体通达交通网，完全覆盖北京东北部区域，车行20分钟可到达国贸、燕莎、望京、丽都、三元桥等都会核心商圈，10分钟可到达首都国际机场。

项目背景

北京是有着三千年历史的国家历史文化名城。北京在历史上曾为六朝都城，在从燕国起的2 000多年里，建造了许多宏伟壮丽的宫廷建筑，使北京成为中国拥有帝王宫殿、园林、庙坛和陵墓数量最多，内容最丰富的城市。北京四合院文化亦是历史久远，底蕴深厚。北京四合院讲究围合，庭院方阔，是人们休闲娱乐的聚集地。北京作为中国的首都，在文化上也是海纳百川，在不断积聚城市文化底蕴的同时，亦在吸纳外来文化的精髓。

规划布局：三街五巷八坊

项目用地分三个部分：即南侧双拼低层住宅用地、中部及东西侧联排低层住宅及合院低层住宅用地、北侧多层住宅及多层公租房住宅用地。三部分建筑自南向北展开，高低错落。

项目溯源福州三坊七巷规制，传承北京四合院的“城市里坊”形制，创新规划“三街五巷八坊”，将中国传统街巷情趣与院落居住进行了完美融合，营造出极具中国境界的诗意坊巷。独院和联排两种户型打造出独具韵味的门第体系，三重庭院体系回归中国人寄情传统的理想人居。

Location Overview

The project is close to three highways—Beijing-Chengde Highway, Beijing-Pinggu Highway and Airport Expressway. In addition, Shunbai Road and Shunhuang Road pass through from south of the site to north of it, and it is also close to Subway Line 15, enjoying a three-dimensional transportation net which covers the northeast area of Beijing. It takes about 20 min drive to capital core business centers such as Beijing International Trade, Yansha Outlets Shopping Center, Wangjing New World Department Store, Lido Place, Sanyuan Bridge, etc., and only 10 min drive to the Capital International Airport.

Project Context

Beijing is a famous historic city over 3,000 years, and it has been the capital city for 6 dynasties in Chinese history. From the State of Yan, many magnificent and splendid royal buildings had made Beijing to be a city with the most abundant imperial palaces, gardens, temples and mausoleums during the past 2,000 years. Beijing quadrangle dwelling has long history and profound cultural foundations. Quadrangle dwelling highlights enclosure and square courtyard layout, and the courtyard is a place for people to enjoy entertainment. As the capital city in China, it welcomes all sorts of cultures. In the process of accumulating itself cultural deposits, Beijing is absorbing external cultural essence.

Planning Layout: Three Streets, Five Lanes and Eight Alleys

The plot is divided into three parts, the first part is the south semi-detached low-rise residential land; the second part is the central, east and west townhouse low-rise residential land and courtyard low-rise residential land; the third part is the north multistorey residential and public housing land. The three parts expand from south to north, in a disordered heights layout.

The project draws on Fuzhou Three Lanes and Seven Alleys layout and inherits Beijing quadrangle courtyard “lanes and alleys” constitute form, and creates a new concept of “three streets, five lanes and eight alleys”, which perfectly integrates Chinese traditional lanes into courtyard dwelling, taking on Chinese flavor poetic residence. The structure of single house and townhouse forms a distinct family status system, while treble courtyard system presents Chinese traditional ideal dwellings.

北京院子

建筑设计

项目运用新中式的建筑风格，演绎当代中国的居住美学。设计在充分借鉴传统元素的基础上采用现代建筑的构造方式，既有中式风格的韵律和层次，又有现代的简洁和时尚质感。灰瓦粉墙、围合布局、前庭后院，还有很多建筑细部都十分注重对传统建筑符号的运用，木色凸花窗、中式屋檐、镶铜雕花木门及砖雕等传统经典装饰，营造出高雅、宁静、舒适的居住氛围。

Architectural Design

The project adopts neo-Chinese style to reveal the aesthetics of Chinese contemporary dwelling. The design fully relies on traditional elements and makes use of modern architectural construction mode, not only containing Chinese rhyme and layered feeling but also interpreting contemporary concise and fashionable flavor. The architectural details lay emphasis on traditional architectural signals such as gray pile, white wall, enclosure layout, atria and backyard, carved wooden window, Chinese eave, brass-insert carved wooden door and brick carving, etc., these classical adornments embellishing an elegant, tranquil and comfort ambiance.

景观绿化

北侧多层住宅组团利用前后两排楼之间空地形成两个集中的块状绿地，布置大尺度的园林景观；南侧低层住宅组团则由每户住宅的南北空地构成分散的点状绿地，各排之间空地形成带状绿地，根据不同的坊巷主题，进行不同的景观小品设计，全冠移植百余种名贵树木，采用天人合一的理念造园。环形景观主轴联系各排住宅宅前绿地。北侧和东侧入口处设计礼仪性绿化景观小广场。

Landscape and Greening

Two focused green blocks, on a vacant space between two rows of the north multistorey residences, form a large-scale garden landscape; sporadic green lots scatter in south and north vacant space of every house of south low-rise residences, and these separate green lots insert between rows of houses, forming ribbon green lands. Various featured landscapes are designed for different themes. Hundreds of rare and valuable trees are transplanted here, creating a garden in unity of heaven and man. Annular landscape axis connects to frontage green lands of every row of houses. There are two little greening landscape squares at the north and east entrances respectively.

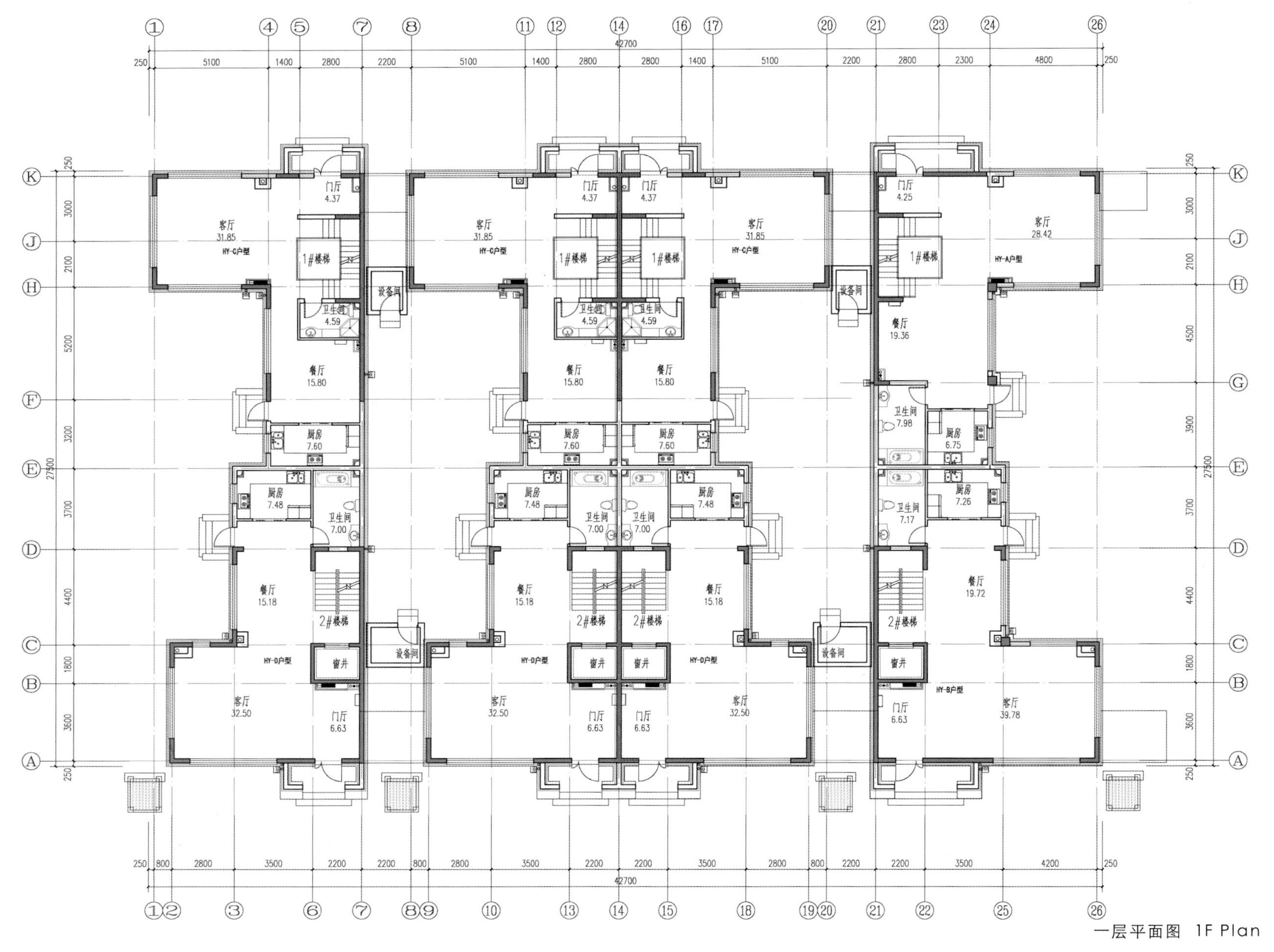

一层平面图 1F Plan

户型设计

北侧多层住宅组团住宅均为一梯两户单元式户型，分为平层户型和跃层户型，户型内部空间设计遵循动静分区、洁污分区、功能空间细化的原则，房间形状方正好用。低层合院住宅分为南北户型，其基本构成为U形，中心为私密室外空间，南北侧为外部空间，内外部空间可以连通，既节省土地又具备基本的低密度住宅品质。

House Type Design

The north multistorey houses are of two units sharing one floor, divided into even layer and leap layer house type. The internal space follows the principles of dynamic and static being apart, clean and dirt being apart, space subdivision and room squaring. The low-rise courtyard residences have south and north house types in U-shape. The central part is for private outdoor spaces, the south and north parts are exterior spaces and these spaces are interconnected. The low-rise courtyard house type is beneficial to save land; furthermore, it has a good quality of low density.

孔子故里儒家理念 传统建筑现代技法

曲阜香格里拉大酒店

Confucian Philosophy in Confucius Hometown   Modern Technology in Traditional Architecture

Shangri-La hotel, Qufu

建筑设计：KKS　室内设计：AB Concept、CCD　项目地址：山东曲阜
总占地面积：50 494平方米　地上建筑面积：53 445平方米　地下建筑面积：2 138平方米　采编：吴孟馨

Architecture Design: KKS　Interior Design: AB Concept, CCD　Location: Qufu, Shandong
Site Area: 50,494 m²　Ground Building Area: 53,445 m²　Underground Building Area: 2,138 m²　Contributing Coordinator: Wu Mengxin

作为孔子的故里，曲阜具有深厚的儒家文化历史底蕴。曲阜香格里拉大酒店地处其中，在设计上将这一文化底蕴以现代的手法融合中国传统建筑文化的精髓展示于人前。青砖飞檐的庑殿式屋顶、红漆回纹屋顶、宫灯式样的超大灯饰、深红色的漆质大屏风等元素不仅展示了传统中式建筑恢弘的气势，亦传达出浓厚的文化气息。而室内设计中所融入的儒家的“礼、德、仁”的三大主题思想以及孔子“礼、乐、射、御、书、数”六艺的理念，更加提升了酒店的文化内涵，与曲阜儒雅的城市气质相呼应。

As the hometown of Confucius, Qufu has profound Confucian historic and cultural foundations. The Shangri-La hotel design aims to utilize modern technologies to display Chinese traditional architectural essence. Traditional Chinese-style elements: hip roof with black brick and overhanging eave, red painted fret roof, super-sized palace lantern-like luminaires and dark red painted screen present imposing momentum and thick cultural breath. The interior design blends in three Confucian main ideas “rite, virtue, and benevolence” and six Confucian classical arts “rite, music, shoot, ride, literacy and math” to elevate cultural connotation and echo Qufu image.

曲阜香格里拉大酒店
La hotel
QUFU

总平面图 Master Plan

区位分析

曲阜香格里拉大酒店位于曲阜市中心，距离孔庙、孔府和孔林仅举步之遥，距离京沪高铁站点仅需15分钟车程。

项目背景

曲阜地处山东省，是中国古代伟大的思想家、教育家、儒家学派创始人孔子的故乡。1982年，就被评为中国首批历史文化名城，而孔庙、孔府、孔林更是被列入世界文化遗产，有着深厚的儒家文化底蕴。曲阜至北京或上海乘高铁只需2~3小时，是京沪两地的短途文化旅行佳处。

规划布局

酒店由2栋大楼——东楼和西楼组成，东楼共设有325间客房和套房，还设有孔咖啡厅，多个开放式厨房全天候奉上国际特色美食；大堂酒廊提供丰富的茶饮。酒店还设有1 600平方米的无柱齐鲁大宴会厅，其前厅连接宽敞明亮的户外露台，此外还有6个多功能厅、一个新娘房和“聚贤堂”贵宾室。

Location Overview

The Shangri-La hotel locates in the downtown of Qufu, just a stone throw to Confucian Temple, Confucius Family Mansion and Confucian Residence. It takes 15 min drive to the Beijing-Shanghai High-speed Rail Station.

Project Context

Qufu is in Shandong Provance, and it is the hometown of Confucius who was a great ideologist, educator and the founder of Confucian school. In 1982, Qufu was designated as the first generation historic and cultural city in China, together with Confucian Temple, Confucius Family Mansion and Confucian Residence were listed as world cultural relics, hence the area has profound Confucian cultural foundations. Only 2 to 3 hours to Beijing or Shanghai by high-speed rail, Qufu is a desirable place for an excursion.

Planning Layout

The hotel consists of an east tower and a west tower, amount to 325 guest rooms and suites. It also equips a coffee house, a few open kitchens supplying international cuisines for 24 hours, and a lobby lounge providing multiple drinks. The Qilu Banquet Hall of 1,600 m² without any pillar, which connects a spacious outdoor terrace. In addition, the hotel has 6 multifunctional halls, a bride room and a VIP Room.

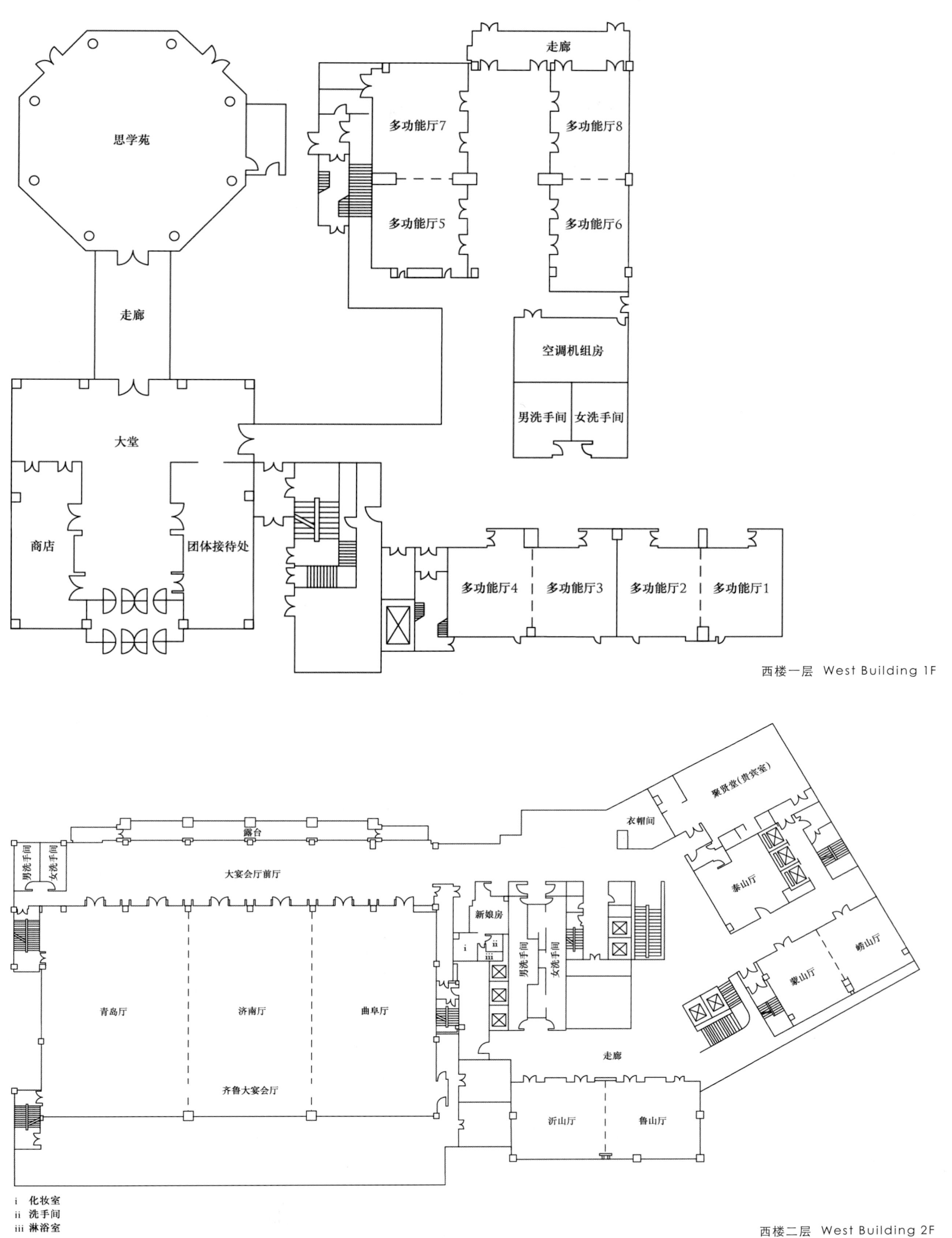

西楼一层 West Building 1F

i 化妆室
ii 洗手间
iii 淋浴室

西楼二层 West Building 2F

建筑设计

酒店巧妙地将儒家理念、中国传统风范及现代建筑风格融为一体，设计以中国建筑特色作为基调，酒店两栋主体建筑正面宛如中式亭苑，青砖飞檐的庑殿式屋顶错落有致。酒店正门车道造型似亭苑，红漆回纹屋顶由八大圆柱支撑，尽显皇家气派。以现代手法彰显历史传承。

Architectural Design

The hotel design integrates Confucian philosophy, Chinese tradition and modern architectural styles. Basing on Chinese architectural features, the front face of the two main towers seem Chinese pavilion with dislocated hip roof covered black brick and overhanging eave. The entrance is like a pavilion with eight roundish pillars holding up a fret painted fret roof, manifesting royal imposing momentum.

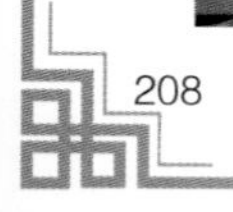

室内设计

酒店的内部设计理念表达了儒家关于“礼、德、仁”的三大主体思想，结合孔子“礼、乐、射、御、书、数”六艺的理念，传递给宾客浓郁的文化气息。

大堂

酒店大堂的设计引申了孔子关于仁德的学说，用建筑和装饰中的层次和对称体现着高雅和谐。大堂顶高8.5米，阳光通透，两棵郁郁葱葱的大树、绿色的小块草坪及深色的木栏勾勒出开阔的庭院结构；宫灯式样的超大灯饰让人眼前一亮；透过落地窗望去，室外中式花园的美景和莺莺绿色展现眼前。深红色的漆质大屏风将空间分割组合，走至大堂的前台，巨幅多媒体完成的百梅图情意盎然。

餐厅

餐厅的内部装饰色彩以柔和的金色、深木色和橙色相呼应，古典而现代。墙面由丝绢装饰，寓意吉祥美满。五个包房和开放就餐区由玻璃墙面隔开，上绘水墨山水花鸟图。香宫深红色木质回纹大门现代风雅，入口两侧的喷泉水池中青铜的锦鲤跃然嬉戏，室内多处使用的水元素令香宫更显灵性。

Interior Design

The interior design shows respect for three main ideas “rite, virtue, and benevolence”. It also integrates Confucian six classical arts “rite, music, shoot, ride, literacy and math” to deliver thick cultural breath.

Lobby

The lobby design borrows Confucian benevolent and virtue concepts, and it uses architectural and ornamental layers and symmetric forms to interpret elegance and harmony. The lobby is 8.5 m high, enjoying opulent sunshine. Two verdant trees, green sods and dark wooden railings outline a broad yard. Palace lantern-like luminaires are eye-catching. Through floor-to-ceiling window, Chinese-style garden with beautiful scenes and lively green leap into sight. Dark red painted screens have a proper partition effect, and when you are close to the reception desk, a multimedia picture of hundreds of plum blossom open right in front of you.

Dining Hall

The interior decoration in dining hall applies gentle gold, dark wood and orange hues, classical and modern elements running parallel. The wall is adorned with tiffany which embroidered with Chinese abstract flowers, birds, yards and pavilions, implicating auspicious luck. Five private dining rooms and open dining area are divided by glass walls with ink landscape painting. Dark red wooden fret door presents elegance, and on both sides of the entrance sits fountain pools with bronze fancy carps leaping and splashing in as well as several waterscapes bring a spiritual feeling to the Shang Palace.

川渝民居

Sichuan-Chongqing Dwellings

四川是多民族、多文化区域，在自然地理条件上有着独特的地理风貌，四川的山地在中国的自然地理上形成了一道独特的风景线。同时，在地理和文化等因素的影响下，也形成了当地独特的建筑文化，而在历史上川渝本就同源，川渝建筑文化亦属于同根而生。而山地建筑、富有民族风情的建筑，是川渝建筑的写照。此外，因为当地生活的闲适与悠闲，在居住方式和邻里关系上建筑又有着田园的诗意。除此，历史因素也是不可忽略的，建筑亦是具有时代性，不同的时代不同的印记，并对当下的建筑文化有着深远的影响。

在此，我们从历史印记、文化传统、民族性及地域自然条件等因素为切入点，诠释建筑的地域特色。

宽窄巷子是成都极具代表性的建筑，呈现了现代人对于城市的记忆，在这里可以体验老成都的原真生活，有着深厚历史背景的宽窄巷子对建筑设计的影响也是极其深远的。如成都钓鱼台精品酒店，它的设计带着设计师对宽窄巷子文化的深入领悟，但为了迎合当下大众群体的心理诉求，设计又融合了法兰西风情，营造出既具有法兰西浪漫风情又具有时光穿越感的入住体验的酒店。

吊脚楼被称为巴楚文化的活化石，是川渝建筑文化的代表。吊脚楼多依山靠河就势而建，为半干栏式建筑，是基于当地地形和自然条件而产生的一种建筑形式。随着社会经济与文化的发展，吊脚楼建筑亦融入了发展的大潮中，融合了新的元素，但仍是川渝建筑的主要特征。如重庆北碚悦榕庄酒店的设计灵感就来源于重庆独特的民国建筑风格和传统的重庆多层民居——吊脚楼；重庆天景山院里的设计则利用重庆传统民居里使用的爬坡、吊脚等适应山地地形的建筑手法，具有极强的重庆房子的特点；缙云山国际温泉度假区的设计亦是以江畔吊脚楼与重庆文化底蕴为源泉。除此，还有将当地的具有象征性的文化特点和景物融入到设计中也是地域特色的体现，如重庆大足香霏古街，将大足县的海棠文化和石刻文化融入到设计之中。

Sichuan has multiple nationalities and multi-cultures, and it owns unique mountainous topography in China. Sichuan forms typical architectural cultures due to geographical and cultural elements. Sichuan and Chongqing had same root in history and culture, so both of them feature in mountainous and ethnic flavor buildings. In this region, local people are used to leading a relaxed life, so habitation and neighborhood appear an idyllic flair. In addition, historical factors cannot be neglected because architecture has characteristics of the times. Different times have their unique architectural characteristics that exert great influences on modern architectural culture.

Here, we try to interpret regional architectural characteristics from the aspects of history, culture tradition, national characters and natural conditions.

Kuan Alley and Zhai Alley is representative buildings in Chengdu. It bears modern people's memories for old city. Here, we can experience a native life in old Chengdu. The Kuan Alley and Zhai Alley has deep historical backgrounds, so it has extremely profound influences on modern architectural design. With in-depth understanding of Kuan Alley and Zhai Alley culture, Diaoyutai Boutique was built. However, in order to cater to contemporary customers' psychological demands, the design intermingles France amorous feelings. The finished hotel boasts French flair, and meanwhile it becomes a space beyond time.

Stilted building is called a living fossil in Bachu culture, and it is also a representative in Sichuan-Chongqing architectural cultures. Stilted building usually builds by rivers and against hills. It is a wooden stem-column building due to regional topography and natural conditions. With the developments of economy and culture, the stilted building is in the tide of social developments. Though new elements are added in, they dominate in Sichuan-Chongqing architecture still. Tianjing Shan Yuan Li takes fully advantages of inclining and stilted methods from traditional Chongqing dwellings, presenting Chongqing folk housing characteristics; Jinyunshan International Hot Spring Resorts is inspired by waterfront stilted buildings and Chongqing cultural foundations; moreover, local cultural symbols and representative landscapes are absorbed in design, likewise, Fragrant Begonia Street integrates Dazu begonia culture and rock carvings culture in its design.

吊脚楼印记 契合自然生长

重庆北碚悦榕庄酒店度假村

Stilted Building Echoing Nature

Beibei Banyan Tree Resort, Chongqing

开发商：重庆柏椿实业有限公司　建筑设计：新加坡悦榕控股集团Architrave Design and Planning　景观设计：意大利迈丘设计事务所
项目地址：重庆市北碚区澄江镇温泉路101号　占地面积：58 566.1平方米　建筑面积：27 160.46平方米　容积率：0.46　绿化率：40.14%　采编：陈惠慧

Developer: Chongqing Baichun Industrial Co. Ltd.　Architecture Design: Architrave Design and Planning　Landscape Design: Metro Studio
Site Area: 58,566.1 m2　Building Area: 27,160.46 m2　Plot Ratio: 0.46　Greening Rate: 40.14%　Contributing Coordinator: Chen Huihui

重庆北培悦榕庄酒店度假村以缙云山、北温泉为自然背景，镌刻其自然生态之美。横竖交错的木柱和横梁，配上青砖灰墙竹板以及架空的木质天花板打造出的吊脚楼建筑与自然相契合，将重庆独有的民居吊脚楼以现代的手法展示人前，呼应巴渝文化及民国阶段的历史文化。结合牡丹和莲花图案以及蜀绣、川剧脸谱和漆器等室内装饰元素，塑造出充满重庆历史人文气息的现代风情度假村。

Beibei Banyan Tree Resort takes the Jinyun Mountain and the North Hot Spring as natural backgrounds to display its ecological aesthetics. Stilted buildings are composed of crisscross wooden pillars and cross beams covering with dark bricks and gray walls and appending with wooden ceilings, well echoing nature. Using modern methods presents Chongqing peculiar stilted dwellings, and meanwhile corresponds with Sichuan and Chongqing culture and Republic of China (ROC) period history. It is a modern resort village full of Chongqing historic and humanistic breath, and you can enjoy peony and lotus graphic patterns, Sichuan opera facial mask and lacquer ware everywhere in public areas and guest rooms.

区位分析

重庆北碚悦榕庄坐落于环境清幽充满历史人文气息的重庆北碚十里温泉城，紧靠重庆风光壮美的国家级自然保护区——缙云山和久负盛名的北温泉，周围依山傍水，宁静祥和。

规划布局

项目由悦榕、悦椿2个酒店以及独栋、双拼别墅和部分公寓组成，分三期开发，分为山地宅院、湿地会馆和森林别墅。项目规划有1套总统别墅、13套豪华别墅、31套标准别墅、18套双卧别墅、48套酒店套房、3~4个专营店、1个SPA温泉中心、1个会议中心。

Location Overview

The Beibei Banyan Tree locates in the Tenmiles Hotspring City and next to national nature conservation area-the Jinyun Mountain and the North Hot Spring. It is surrounded by mountains and rivers, quiet and peaceful.

Planning Layout

The project includes the Banyan Tree Hotel, the Angsana Hotel, detached and semi-detached villas and some apartments. The whole project is divided into three phases, and they are mountain residences, wetland clubs and forest villas. The project is planned with a presidential villa, 13 luxury villas, 31 standard villas, 18 double bedroom villas, 48 hotel suites, 3 or 4 exclusive stores, a hot spring SPA center and a convention center.

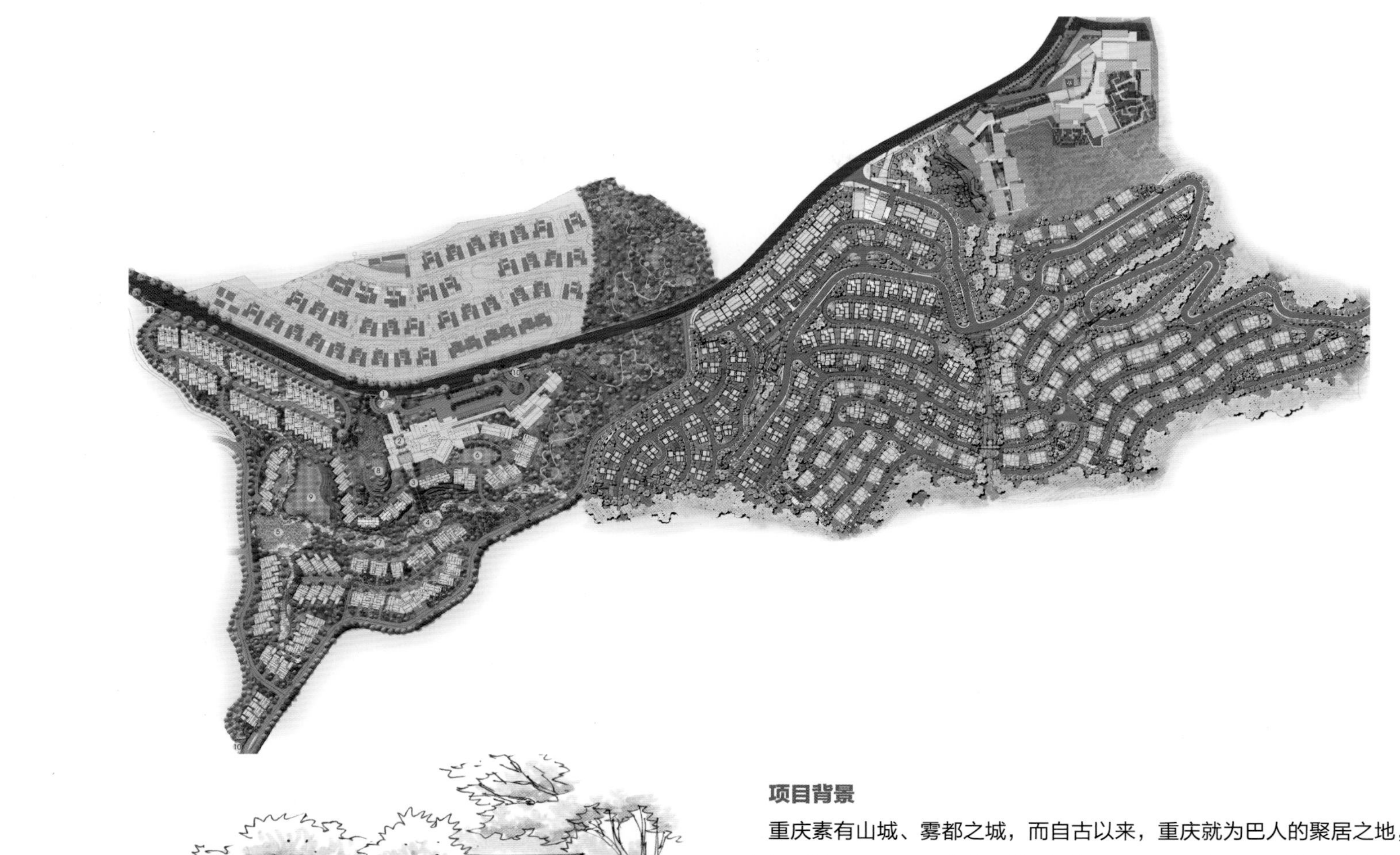

项目背景

重庆素有山城、雾都之城，而自古以来，重庆就为巴人的聚居之地，但因地势刚险，巴人习惯背倚山川、逐水而居，亦因此，吊脚楼由此产生。吊脚楼为半干栏式建筑，依地形屹立，依山就势，高低错落，有层层出挑之感。在绿影婆娑中傲视山水，与自然和谐统一。随着当地经济文化的发展，吊脚楼建筑作为地方民居，也在不断地演变发展，但其淳朴自然、散落于山水间的特点却也更为适合人们所追求的诗意栖居的居住理念。

在重庆的发展历史上，也曾受殖民侵略的影响，在建筑文化上，涌入了西方的建筑文化，在发展的过程中不断地与当地文化相结合，形成中西合璧的建筑文化。在民国时期，重庆作为陪都，民国建筑给重庆留下了不可磨灭的历史烙印，如以民国历史、巴渝特色为主的重庆特色的民国街，不仅重现了民国时期重庆的著名建筑和景观，更成为了人们走进历史的捷径。

Project Context

Chongqing is a mountain city and well known for mist. Since ancient times, Sichuan and Chongqing people have settled here, and due to local precipitous topography, it formed the settlement customs of leaning against mountains or standing up on waters, so stilted building was generated. The stilted building is a wooden stem-column building, and it builds timber pile in accordance with the terrain level, forming a layer-and-layer progressive feeling. These stilted buildings ensconce in mountain with plentiful landscape resources and harmonize with nature. With the development of economy, the stilted buildings as vernacular dwellings have developed as well. It features in pristine and nature, interspersing in landscapes at random, which becomes a poetic residing concept for modern people to pursue.

Chongqing used to be a western colonial city, so western architecture was packed in in colonial period. With the development of history, the western architecture gradually integrated in local architecture, forming a combination of Chinese and Western architectural culture. Chongqing was an auxiliary capital in Republic of China period, therefore ROC style architecture left indelible brands. Such as the ROC Street, boasting ROC style history and Sichuan and Chongqing characteristics, reappears famous buildings and landscapes of ROC period, and it also acts as an accessible road leading people back to history.

建筑设计

酒店的设计重点突出了川渝地区的审美观念，并将当地文化特色嵌入其中，以民国时期的建筑风格为主题，以横竖交错的木柱和横梁，配上青砖灰墙竹板以及架空的木质天花板，呈现重庆独有的多层民居——吊脚楼的建筑特色。同时，融合重庆山水风光和巴渝文化特色，结合缙云山的地形地貌，萃取“外封闭，内开敞，大出檐，小天井，高勒脚，冷摊瓦”的建筑元素，完美融合现代元素，将建筑巧妙地融入缙云山连绵的画卷之中。采用巴渝文化的酒店整体建筑风格，内部开放式的空间格局不仅再现了灵活多变的建筑精神，也将生活空间与天地景色融合为一。

Architectural Design

The design highlights an aesthetic sense of Sichuan-Chongqing district and combines with local cultural characteristics. Themed with ROC period architectural style, the buildings are composed of crisscross wooden pillars and cross beams. Wrapping with black bricks and gray walls and appending with wooden ceilings, Chongqing characteristic multilayered dwellings come forth-stilted buildings. In virtue of beautiful scenery and cultural features of Sichuan and Chongqing, the stilted buildings combine with topography of the Jinyun Mountain and abstract architectural elements of “outside close, inside open, long eave, small patio, high plinth and cold-colored tiles” to perfectly integrate with modern elements.

小贴士

小青瓦、坡屋顶、白灰墙、雕花窗等虽是重庆传统民居的主要建筑元素，但爬坡上坎亦是山城重庆的一大特点，而基于地形而建的建筑，也形成了爬坡式的建筑形式，形成高低错落有致的建筑排布，使建筑很好地融入到自然之中，并具有良好的景观视野。

Tips

Black tile, slope roof, white and gray wall and lattice window are main building elements in Chongqing traditional dwellings, while slope architecture is indeed first noticeable feature in Chongqing architecture. Buildings are arranged according to landforms in dislocated configuration, so buildings well blend in nature and enjoy good landscape vision.

景观设计

景观设计沿袭了悦榕庄营造优雅浪漫的度假风格，同时融入原生态的文化元素。风格采用巴蜀文化中的吊脚楼设计元素，结合现代设计手法，营造清幽又豪华的气氛。大量取材当地竹子，藏奢华于隐处，将整个巴渝建筑风格的酒店掩映在廊桥翠竹之中。

硬景设计将现代与传统元素相结合，营造人与自然和谐共生的绝佳体验，精巧构思，浑然天成，处处彰显巴蜀文化。并将场地用特色铺装连接起来，统一整体空间。河流靠近别墅群，使别墅亲水，并加设木平台，整体提升别墅的价值。此外还加设快速通道，使行人能更便捷地通行。软景设计实现“内饰四季”的效果，充分利用当地植物与自然水源，引以“瑶池美景，奇花异草”的概念，结合高差和台地，利用不同植被的高度包囊硬质景观，形成特色的梯田式植被群落。

Landscape Design

The landscape design follows elegant and romantic holiday style of Banyan Tree and integrates in native cultural elements. It combines stilted building design with modern design approach to create a quiet and luxury atmosphere. The design uses a large number of local bamboos to decorate the Sichuan and Chongqing architectural style hotel.

Hard landscaping perfectly combines modern and traditional elements to create a brilliant experience of harmonious co-existence of human and nature. The ingenious idea is natural to manifest Sichuan and Chongqing culture at every corner. The design connects up different parts by featured pavements into a unity. The villa group is close to river and equips with wooden platforms, enhancing the value of the villas integrally. In addition, there are fast tracks for customers enjoying a more convenient walking. Soft Landscaping makes full use of local plants and natural water source to realize four seasons scenery, and meanwhile it adopts a concept of immortal abode with exotic flowers and herbs to combine discrepancy in elevation and platforms and wrap hard landscapes by different plants, creating a featured terraced-field plant group.

BANYAN TREE
悦榕私邸
—山地宅院—

室内设计

项目室内装饰与建筑设计相呼应，融入了各种传统重庆风格的设计元素。度假村随处可见牡丹、莲花图案，前者象征着高贵、财富和好运，而后者代表幸福。酒店的公共区域及房间的内饰都采用蜀绣、川剧脸谱和漆器等装饰，演绎出中式的传统典雅。

Interior Design

Interior design integrates in elements of Chongqing traditional architecture style. Peony and lotus graphic patterns can be seen everywhere in the resort, because peony symbolizes noble, fortune and good luck while lotus stands for happiness. Sichuan embroidery works, Sichuan opera facial masks and lacquer wares appear in public areas and guest rooms frequently, embodying a Chinese traditional elegance.

巴渝村落模式 山居聚落空间

重庆天景•山院里

Sichuan and Chongqing Village Mode Mountain Settlement Space

Tianjing Shan Yuan Li, Chongqing

开发商：重庆天景房地产置业有限公司 建筑设计：重庆博建建筑设计有限公司 景观设计：重庆日清国际景观设计有限公司 项目地址：重庆南岸区学府大道69号
占地面积：27 000平方米 建筑面积：19 000平方米 容积率：0.7 绿化率：37%
设计团队：侯宝石、龙涛江、张晴波、王强 摄影：陈本均 侯宝石 采编：谭杰

Developer: Chongqing Tianjing Real Estate Ltd. Architecture Design: BOA Design Landscape Design: La Cime Internationale
Location: No.69 Xuefu Avenue, Nan'an District, Chongqing Site Area: 27,000 m² Building Area: 19,000 m² Plot Ratio: 0.7 Greening Rate: 37%
Design Team: Hou Baoshi, Long Taojiang, Zhang Qingbo, Wang Qiang Photographer: Chen Benjun, Hou Baoshi Contributing Coordinator: Tan Jie

地势的高差与坡度为构建山地建筑提供了前提，也是重庆典型的地形。因此，项目设计因势利导，平行于等高线布局别墅，采用巴渝地区常见的村落古街模式，利用重庆传统民居里使用的爬坡、吊脚等适应山地地形的建筑手法，配以传统白墙与褐红色砖的运用以及木材装饰的点缀，使建筑拔地而起，因地而生，结合退台和跃层的设计，将重庆的山地建筑的优势演绎得淋漓尽致，同时也将重庆独有的吊脚楼的风韵细腻地表现出来。在传承当地传统建筑文化的同时，采用现代材料和技术，如钢方管的椽条，在传统中，透露着现代气息。

Topography discrepancy and gradient are prerequisite for mountainous buildings, which are typical landforms in Chongqing. In accordance with local terrain, villas ensconce along isohypse lines. The layout adopts ancient village mode, common seen in Sichuan and Chongqing. Chongqing traditional dwelling construction techniques, such as inclining and stilted ways, adapt to mountain topography well. Traditional white walls, brown red bricks and wooden ornaments make buildings rising straight from the ground. Retreating and spring layer methods fully display the advantages of mountainous building in Chongqing, meanwhile peculiar flavor of stilted building is exhibited. The project inherits local traditional architecture culture and adopts modern materials and technologies, such as square steel tube rafter, integrating a modern ambiance in tradition.

住宅总平面图 Residence Master Plan

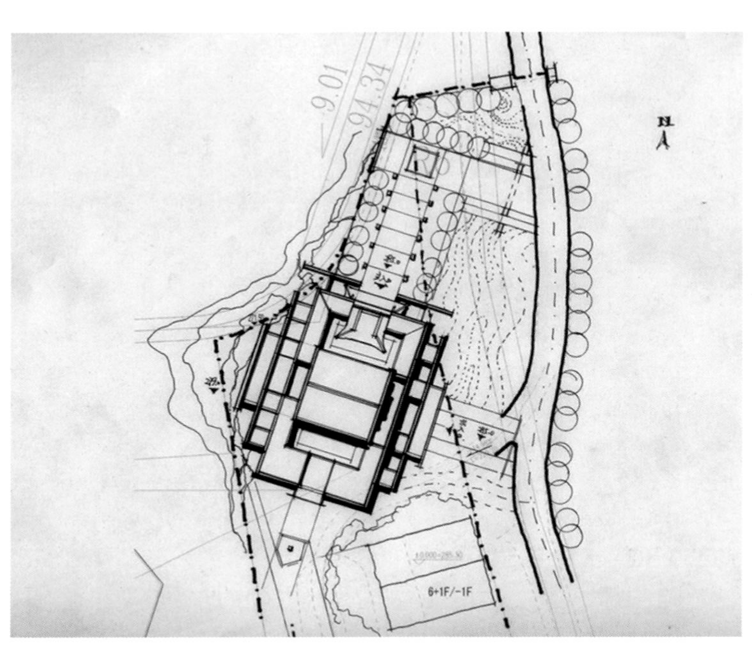

会所总平面图 Club Master Plan

区位分析图 Location Map

区位分析

项目位于重庆市主城区西南部真武山西麓南山脚下，东靠真武山原生保护林带，西望重庆交通大学校园。

项目背景

城市背景

重庆素有山城之称，山地建筑是重庆特有的建筑形态之一。山地建筑因势利导，与自然相契合。而吊脚楼的出现亦是因地势而来，是重庆山地建筑的特征之一。吊脚楼倚山逐水，散落在青山绿水间，带着自然淳朴的乡土气息，成为了重庆独特的建筑文化和标识。

地势分析

项目地块呈南北狭长形态，地形复杂，西低东高，最低点和最高点绝对标高分别为249米和306米，高差达57米，基地范围80%以上区域场地坡度均为30%以上，是一个典型的山地项目。

设计构思

中国文化聚落：南山自古多文人逸士居住，采菊东篱，悠见南山，再现中国传统式山居聚落空间，中国式街道村落的特点，用中国式的建筑塑造中国式的生活。

尊重自然利用坡地：保存原来地貌特点的前提下，利用地形的高差设计高低错落，形成带有跃层和吊层的典型山地建筑。

Location Overview

The project locates at the foot of the Nanshan Mountain, the west of the Zhenwu Mountain, the southwest of central downtown, Chongqing. It is adjacent to the east of the Zhenwu Mountain Original Forest Protection Zone, and westward overlooks Chongqing Jiaotong University.

Project Context

City Background

Chongqing is called a Mountain City, hence mountainous building is one of unique architecture in Chongqing. Mountainous buildings construct in accordance with local terrain and echo nature. Stilted building just generates according to topography, which is one of features of mountainous building. Stilted buildings settle against mountain and beside water, scattering among landscapes and bringing natural pristine flavor. Stilted buildings become Chongqing peculiar architectural culture and symbol.

Topography Analysis

The plot is in a narrow rectangular form along south to north with complicated terrain descending from east to west. The highest and lowest absolute altitudes are 306 m and 249 m respectively with height discrepancy of 57 m. In the plot, 80% site is sloping fields with gradient over 30°. It is a typical mountainous project.

Design Conception

Chinese Cultural Settlement

The Nanshan Mountain has been a famous solitude attracting many literatus from ancient times. The ancient poem says: picking chrysanthemums under the eastern fence, and leisurely see the Nanshan Mountain. The project restores Chinese traditional mountain settlement spaces and Chinese-style village features. Using Chinese architecture makes Chinese life style.

Respecting Nature and Utilizing Sloping Fields

Under the original terrain and special conditions, the design utilizes height discrepancy to arrange buildings' levels, forming mountainous buildings in spring layer and hanging layer forms.

规划设计：山地场镇布局

项目由13栋联排别墅、1个会所、1个车库组成，虽然不大，但是处于南山大环境下，总量小而有神韵。

项目设计因势利导，平行于等高线布置别墅，形成川渝传统山地场镇的布局格调，并以会所作为小区入口必经标识，有四川罗城古镇的韵味。规划布局采用巴渝地区常见的村落古街模式，沿平行等高线延展出的长街，活动和交往都在街上形成。由于高差，一面是爬坡向上的建筑，另一面则是跌落向下的建筑，在建筑之间设计景观和绿化，从而可以仰观山，俯瞰景。会所采用川东传统山地院落的轴线和空间布局，串起整个项目的生活环境，既适应山地又颇具气势，使得整个村落显得勃勃生机。

Planning Design: Mountain Town Layout

The project consists of 13 townhouses, a club and a car park. Although it is small, under the environment of the Nanshan Mountain, it appears ingenious charm.

In accordance with local terrain, townhouses ensconce along isohypse lines, forming Sichuan-Chongqing featured mountain town layout. The club is an unavoidable indicating building at the entrance of the community. The planning layout adopts ancient village mode, common seen in Sichuan and Chongqing. Long streets, extending with parallel isohypse lines, are ideal activity and communication places. Due to the height discrepancies, there are ascending buildings and downward buildings, and between them are landscapes and greening, so you can looking up the mountain and overlook the landscapes in the same place. The club uses eastern Sichuan tradition: using space layout and mountain courtyard axis to string up a whole living environment, which complies with hilly area and enjoys imposing manner in the mean time, making the whole village vibrantly.

山院里
21-1

建筑设计：现代版重庆民居

设计借用传统中式的建筑风格的元素，利用重庆传统民居里使用的爬坡、吊脚等适应山地地形的建筑手法，具有极强的重庆房子的特点。并结合了坡屋顶、廊道、马头墙、丰富的屋面轮廓等元素，强调建筑形体多变的组合，利用退台、跃层等多种空间和形体组织方法形成多种层次的室外平台和阳台；端户型打破传统的山墙形象，使建筑在各个方向都互为景观。会所采用典雅的中国式祠堂的设计风格，注重比例与尺度，注意砖石材等材料的选择，营造传统聚落的精神家园。

在材料和主次空间的链接上，运用现代材料和技术手段以及简化抽象的内容，如住宅部分的抽象变形的挑檐拱架，会所墙体的现代构成式的平行线条，钢方管抽象的椽条等，使得整体建筑群既本土化，又具有现代气息。通过使用传统的白墙和来源于泥土的褐红色砖，使建筑和环境融为一体，局部点缀以木材的外装饰，突出了“家”的温暖和感觉。

Architectural Design: Modern Version Chongqing Dwellings

Architectural design uses traditional Chinese style elements for reference and adopts Chongqing traditional dwelling construction techniques, such as inclining and stilted ways that adapt to mountain topography, so the buildings have the characteristics of Chongqing dwellings. Sloping roof, gallery, wharf wall, multiple roof profile form multiple architectural form combination. Retreating and spring layer spatial organization methods extend outdoor terrace and balcony. Side house type breaks through the image of traditional gable wall, making every side serve as a foil to one another. The club employs an elegant Chinese ancestral hall style, highlighting proportion and scale, maximizing brick and timber materials, and building a spiritual homeland of traditional settlement.

The design fully uses modern materials and techniques, and emphasizes the connection of primary and secondary spaces. It highlights simplified and abstract context, such as abstract deformation overhanging eave and arch center in residence and modern parallel lines and square steel tube abstracting rafter on club walls. The design embraces localization and modernism. Traditional white walls and brown red bricks made of local mud integrate the buildings in the environment. Wooden external ornaments at parts create a warm homeland.

小贴士

巴渝建筑风格的形成主要源于三个因素：一是自然因素，重庆江水环抱、山势起伏的地形环境和潮湿多雨的气候环境；二是外来移民带来的建筑文化以及西方文化的进入和影响。这些在巴渝建筑风格的发展过程中都留下了历史印记，而现在的巴渝建筑既有着传统建筑元素，又有着现代元素的融入，如传统的白墙、本地的褐色石砖及木材等的应用，都保留着传统的印记。

Tips

The formation of Sichuan and Chongqing architecture style is due to three factors: natural conditions of surrounded rivers, undulating mountainous topography and humid climate; immigrant architectural culture; and western culture invasion. These factors have left historic stamps in the process of architecture developments. Contemporary Sichuan and Chongqing architecture contains traditional and modern elements, such as white wall, brown stone tile and timber.

景观设计：与自然相生

“因地制宜，显山透绿”是景观设计的设计理念，这里不是要重造人工环境，而是把南山的自然环境延续下来，使建筑置身于自然之中。

Landscape Design: Coexisting with Nature

Conforming to topography and exhibiting landscapes are design concept of the project. Designers try to extend the Nanshan Mountain natural environment and set buildings in natural environment instead of restoring an artificial environment.

户型设计

鉴于道路平行于等高线布置，因此垂直于等高线的两侧住宅建筑以爬坡、吊脚的方式设计户型，同时设置立体院落，即前后院分别在不同标高位置，甚至在边户型形成前、侧、后三个标高的院落组合。吊层结合设备用房和车库使用，节省用地。

House Type Design

Due to roads are parallel with isohypse lines, the residences, perpendicular in two sides, adopt inclining and stilted patterns as house type. Designers innovate three-dimensional courtyards: the front and rear yards are on different isohypse positions, and there is a courtyard combination of front, side and rear yards in different layers seen from side. The spring layer structure uniting with spare rooms and garages vacates much space.

宽窄巷子文化底蕴　法式浪漫巴适生活
成都钓鱼台精品酒店
Kuan Alley and Zhai Alley Cultural Foundation　French Romantic and Comfortable Life
The Diaoyutai Boutique, Chengdu

建筑改造/室内设计：4BI建筑事务所　设计师：Bruno Moinard
项目地址：成都宽窄巷子　项目面积：约16 000平方米　采编：陈惠慧

Architecture Renovation/Interior Design: 4BI Architecture Firm　Designer: Bruno Moinard
Location: Kuan Alley and Zhai Alley, Chengdu　Site Area: about 16,000 m²　Contributing Coordinator: Chen Huihui

具有300多年历史的都市风情街区——成都宽窄巷子是成都的一张名片，是人们走进成都，了解成都当地生活和建筑文化不可或缺的素材。酒店位于其中，即使融入现代的设计手法和元素，亦不能割舍掉这份浓重的历史文化气息。因此，酒店设计保留了宽窄巷子里老建筑的中式古典神韵，将宽窄巷子300多年的人文气息与钓鱼台品牌内蕴有机结合，同时又融法兰西的浪漫风情于典雅幽静的东方庭院中，彰显出精致华美的同时，又不失摩登风范，演绎现代巴适生活。

Chengdu Kuan Alley and Zhai Alley is a urban style street in Chengdu over 300 years. Now it becomes a name card for Chengdu, and it is an indispensible source for understanding local life and architectural culture. The hotel design adopts modern technologies and elements, while it can not give up a profound historical and cultural ambiance. Hence, the design retains a Chinese classic flair in old buildings of Kuan Alley and Zhai Alley, and integrates its humanistic breath in Diaoyutai brand connotation as well as blends French romantic flavor in oriental court to manifest exquisite and modern styles and display fashionable and comfortable life.

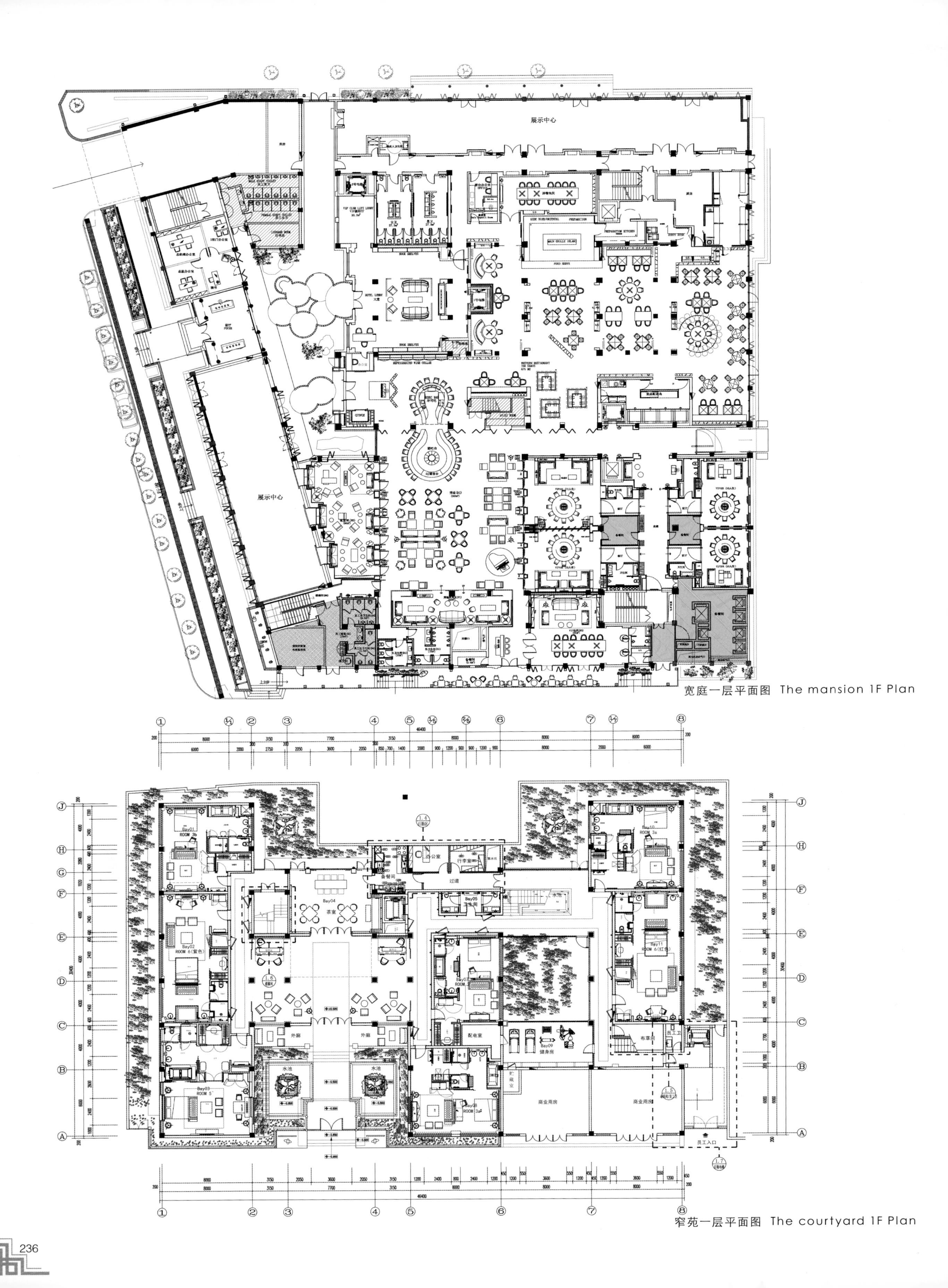

宽庭一层平面图 The mansion 1F Plan

窄苑一层平面图 The courtyard 1F Plan

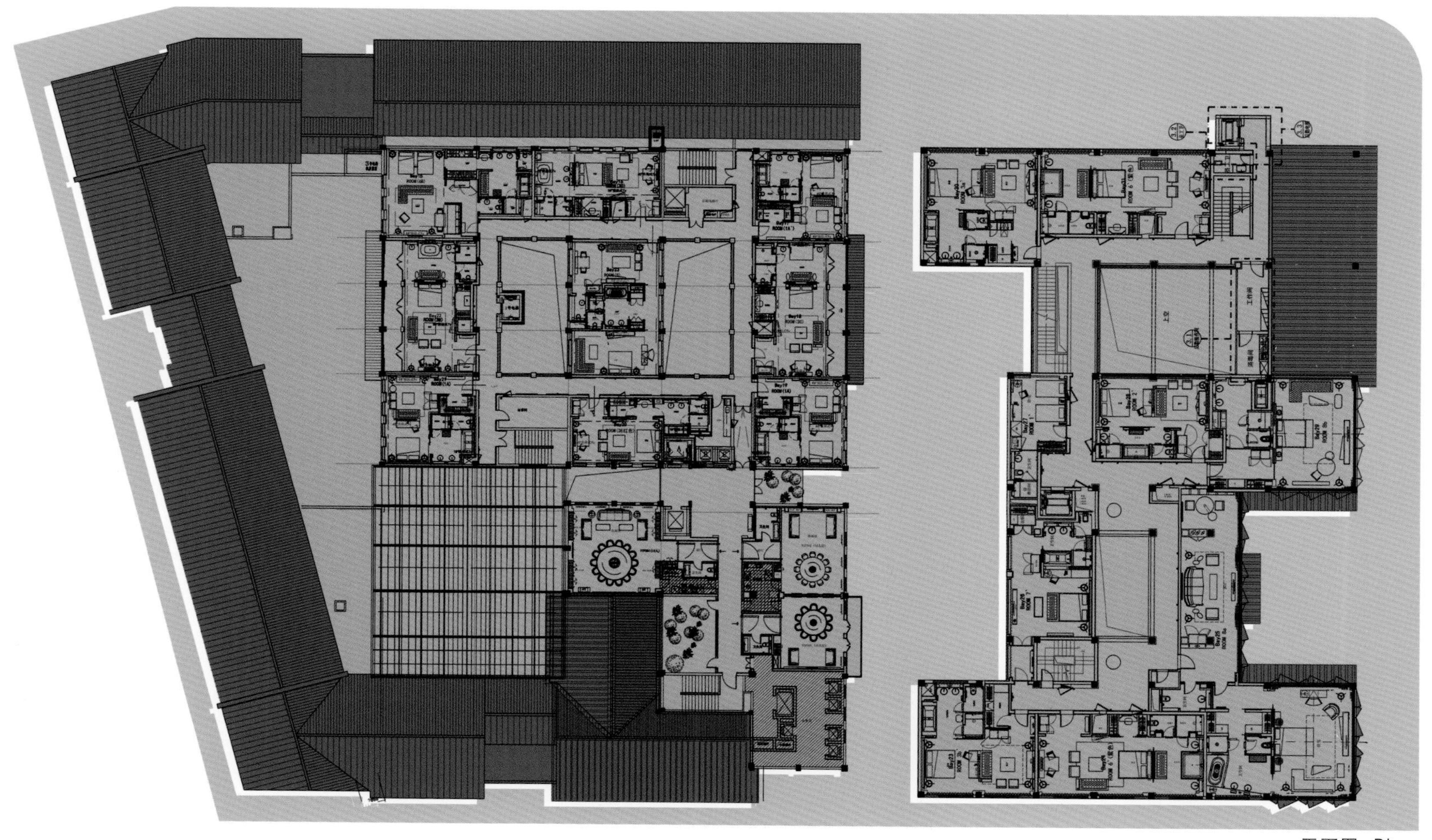

平面图 Plan

区位分析

成都钓鱼台精品酒店坐落于具有300多年历史的都市风情街区——成都宽窄巷子。

项目背景

宽窄巷子是成都的历史文化地标，它是成都超过2 300年建城历史的一个重要片段，由3条有300年历史的古老街道，45座清末民初风格的四合院、花园洋楼等建筑群落组成。灰砖、粉墙、石板路传达出丰富的历史信息，6～8米宽的街道尺度，“鱼骨状”的空间肌理，塑造出传统建筑界面的层次关系，至今保持着老成都市民的生活场景。

规划设计

项目主体由“宽庭”“窄苑”两重院落构成，设有45间极具特色的客房，有着带有中国特色的庭院，融汇当今国宾馆的雍容气度，酒店装饰也融入丰富多样的中国元素，住客来到这里会觉得非常亲切。此外，项目还设有御苑、乐庭-亚洲风尚餐厅、芳菲秀-大堂酒廊、钓鱼台俱乐部、钓鱼台生活馆等配套功能空间。

Location Overview

Chengdu Kuan Alley and Zhai Alley, which the hotel is situated in, is a style street with a history of more than 300 years.

Project Context

Kuan Alley and Zhai Alley is a historical and cultural landmark of Chengdu. It is a significant part of Chengdu's over 2,300 years long history. It is composed of 3 ancient streets all of 300 years old, as well as 45 quadrangle dwellings and garden houses that feature the style of the late Qing Dynasty and the early Ming Dynasty. The gray brick, white wall and slabstone road convey out rich historical signals. The 6-8 m width and the herringbone layout expose the hierarchy of traditional buildings. The lifestyle of old Chengdu has been passing down here.

Planning Design

The hotel highlights two courtyards—The Mansion and The Courtyard, and includes 45 featured guest rooms, as well as Chinese style courtyards, sharing the luxuriousness of a modern state guest hotel. The space is decorated with various Chinese elements, which makes guests feel welcomed upon their arrival. There equips other supporting functional spaces, such as Royal Court Restaurant, Leting Court (Asia Fashion Dining Hall), Fangfei Garden (Lobby Lounge), Diaoyutai Club and Diaoyutai Life House.

建筑设计

建筑设计保留了老建筑的中式神韵，古朴的灰砖凝聚了当地的历史气息。屋顶的优雅弧线以及屋檐与门窗的中式细节都体现出了地域特色与独特的人文情怀。部分客房的屋顶沿着原有屋顶的形状被设计成波浪形，增加了房间的高度，让客人的视野更开阔。每个房间都保留原有的中式窗户，增加酒店的古典韵味。酒店还将法兰西的浪漫风情注入典雅幽静的东方庭院，将宽窄巷子300多年的人文气息与钓鱼台品牌的高贵底蕴相融合，打造出精致优雅，又不失摩登风范的空间。

Architectural Design

The hotel is composed of old buildings, flaunting strong Chinese charm with gray brick drawing in the local historical breath. The graceful arch on the roof, and the Chinese elements on the eave, door and window all reveal distinctive local features and peculiar humanistic feelings. Parts of guest room roof remaining the original shape is designed in a wavy pattern, increasing the height of the rooms and making a broader view for guests. Chinese style windows are retained as well, strengthening classic feelings. Moreover, French romantic flair is infused into the tranquil Oriental courtyard, combining the time-honored cultural touch of Kuan Alley and Zhai Alley with the noble connotation of Diaoyutai. It is an exquisite and graceful, yet modern space.

室内设计

45间客房共16种房型，4个主题色，均散发着东方庭院的文化气息。4间餐厅酒廊设计各异。其中，御苑的设计上，设计师将传统的中国元素，在餐厅空间里进行了国际化的表达，从木质拱形门通过即到达宽敞的餐厅空间。屋顶沿着原有建筑物屋顶的形状用天然的木材装饰，搭配古朴简单的吊灯，和舒适的座椅，细节之处尽显餐厅的高端，大气，与“钓鱼台”一直遵循的品牌定位高度契合。乐庭餐厅为亚洲风尚餐厅，设计师采用新中式融合简约现代的手法，打造一个轻快、自由的用餐空间，细微之处透露着设计师对当地文化的独特理解。芳菲秀即酒店的大堂酒廊，具有一廊三吧（雪茄吧、香槟吧和寿司吧）的功能组合，设计师又赋予了这个空间美妙的灵魂。大堂酒廊的天幕设计，被赋予一种闲适的愉悦感。香槟吧有连通着窄巷子的户外阳台，尤其适合举办20人的朋友聚会。

Interior Design

The 45 guest rooms are classified into 16 types and 4 themes. Guests will feel an oriental courtyard breath upon their arrival. In the Royal Court, traditional Chinese elements are internationalized. The spacious restaurant space can be reached through the wooden arc door, with its ceiling decorated with natural wood, and simple yet vintage chandeliers and cozy chairs. Every detail manifests an upscale taste, which echoes the brand position of Diaoyutai. Leting Court is an Asia Fashion Dining Hall that is a bright and free dining space designed in neo-Chinese and minimal style, and the detail design displays designers' unique cultural understanding. Fangfei Garden is a lobby lounge, dividing into three bars: Cigar Bar, Champagne Bar and Sushi Bar. The ceiling backdrop design brings a relaxed joviality. In the Champagne Bar, there is an outdoor balcony connects Zhai Alley, which is an ideal place for a party available for 20 people.

追寻大足印记 创新中式街区

重庆大足香霏古街

Searching for Dazu Stamp Innovating Chinese Style Street

Fragrant Begonia Street, Chongqing

开发商：泽京集团 建筑设计：DC国际建筑设计事务所 项目地址：重庆大足
用地面积：30 000平方米 建筑面积：19 000平方米 采编：赵俊芳

Developer: Zengjing Group Architecture Design: DC Alliance Location: Dazu District, Chongqing
Site Area: 30,000 m² Building Area: 19,000 m² Contributing Coordinator: Zhao Junfang

建筑作为城市的一部分，同时也是当地文化的一种表现方式。大足素有“海棠香国”之称，石刻文化历史悠久。而走在香霏古街不仅可以看到海棠印记，还可以在广场上看到石刻地雕，将大足的文化特征以具象的方式传达给世人。此外，镂空砖、木色格栅、花格门窗、印花玻璃等丰富的形式和材料以及局部“留白”处理的立面设计，在传承传统中式元素的同时，又不失现代的简约。内庭院、水院、滨水平台、内游廊等景观空间处理手法的运用，又赋予建筑一种更自然、更现代、更具生命力的气质，使项目成为大足的一部分的同时，又带给大足新鲜的气息。

Architecture is a component of a city, and meanwhile it is a symbol presenting local culture. Dazu County is also called “Begonia Aroma Country”, with time-renowned rock carving culture. Walking on the Fragrant Begonia Street, you can see the begonia figures exposing everywhere and rock carving ground statues standing on squares, which embodies Dazu featured culture to the world. Hollow brick, wooden trellis, lattice door and window and printing glass boast opulent facade and leaving blank techniques which present the inheritance for traditional Chinese elements but also appear modern minimalism. Internal yard, waterscape court, waterfront terrace and gallery adopt featured landscape space design techniques to create a natural, modern and vital breath. The Fragrant Begonia Street becomes a part in Dazu, and it injects fresh blood into Dazu at the same time.

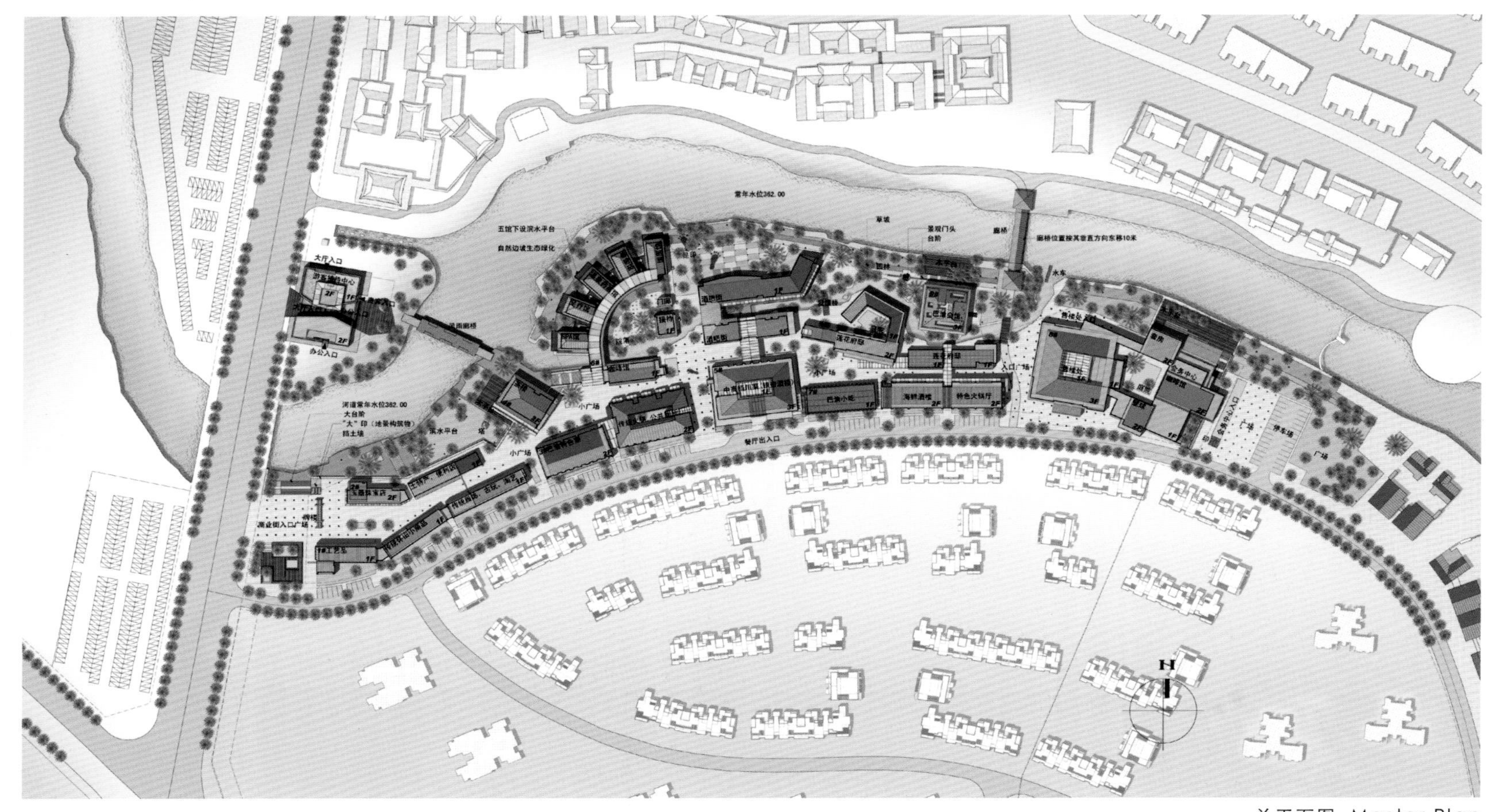

总平面图 Master Plan

建筑及视线分析 Architecture and View Analysis

区位分析

项目位于大足县政府东侧，距县中心约1公里，交通通达性良好，植被丰富。

项目背景

重庆大足县位于四川盆地东南部，重庆市西北部，距离重庆市80公里。大足素有“海棠香国”之称，是世界文化遗产——大足石刻坐落之处，并被国家文化部命名为“中国民间文化艺术之乡”。

项目定位

项目设计定位为海棠香国大足文化本身，融合大足的“海棠文化”和“石刻文化”，不仅仅在设计中运用“海棠”的表面形态，还将融入“海棠”的文化理念；同时，把石刻文化融入到设计中来，在广场设计地雕，从而提升商业街整体的品味形象，丰富视觉景观，使其在将来成为大足石刻文化中的一部分。

Location Overview

The project sits in the east of the Dazu County Government. It is 1 km away from county center with convenient transportation and dense vegetation.

Project Context

Dazu County locates in the southeast of the Sichuan Basin, the northwest of Chongqing with distance of 80 km. Dazu County is known as “Begonia Aroma Country”, where the world cultural heritage-Dazu Rock Carvings exist. The county is designated as the Hometown of Chinese Folk Culture and Art by the State Ministry of Culture.

Positioning

The project relies on the Begonia Aroma Country to combine Begonia culture and Rock Carvings culture. It applies begonia external shape and integrates begonia culture and rock carving in design. The square ground statues improve total taste and image for the commercial street and enrich visual landscapes, and they will become a part of Dazu Rock Carvings culture in the future.

规划设计

项目意在打造为创意新颖、主题突出、技术合理与自然环境优美的现代中式商业街，设有手工艺品销售、餐饮、休闲娱乐、会务四大区域，根据场景的差异策划出不同的商业业态，使之各具特色，并设计通过各个中心节点串联在一起，通过这样分散而又集中的布局带来商业街外场地空间的最大自由度和可能性，打造不同的空间体验。

动线设计通过点、线、面来创造节奏，把购物的街道曲折化。曲折形状的街道在视觉上有鼓励和吸引行人往前探索的作用，再加上广场空间、模糊空间的引力作用，人们行走路径在猎奇心理引导下会自然产生一条生动的商业动线。

Planning Design

The project aims to construct a commercial street with original creation, prominent theme, appropriate technologies and beautiful natural environment. Here are four sections: handiwork sales area, F&B area, leisure and entertainment area and conference center. According to different settings arranges various commercial formats that own typical characteristics. The four areas are connected by central nodes, forming scattered while concentrated layout so that the outside space of the commercial street has adequate place to create different spatial experiences.

The design controls rhythms by smartly using point, line and plane to zigzag the shopping street. A zigzag street is conducive to encourage and absorb tourists to further explore forward. Square spaces and indistinct spaces can guide tourists stepping forward onto a natural business pedestrian circulation under curiosity.

香霏街建筑鸟瞰效果
Fragrant Begonia Street Aerial Rendering

昌州古城建筑效果城隍庙内街
Changzhou Ancient Town Rendering — Chenghuang Temple Street

昌州古城建筑效果 Changzhou Ancient Town Rendering

香霏街建筑效果主入口 Fragrant Begonia Street Main Entrance Rendering

香霏街建筑效果主入口区 Fragrant Begonia Street Main Entrance Area

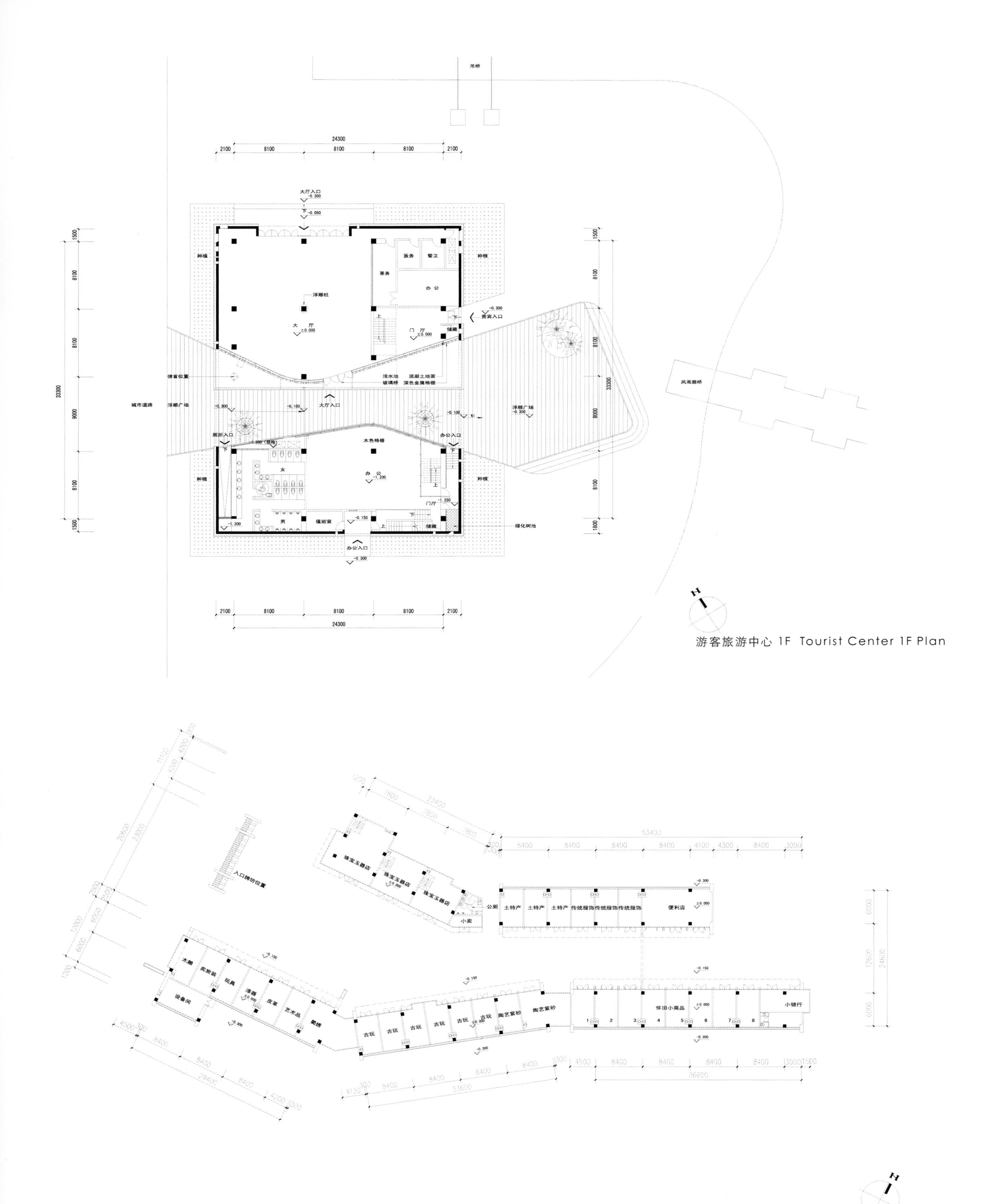

游客旅游中心 1F Tourist Center 1F Plan

入口商业 1F Commercial Street Etrance 1F Plan

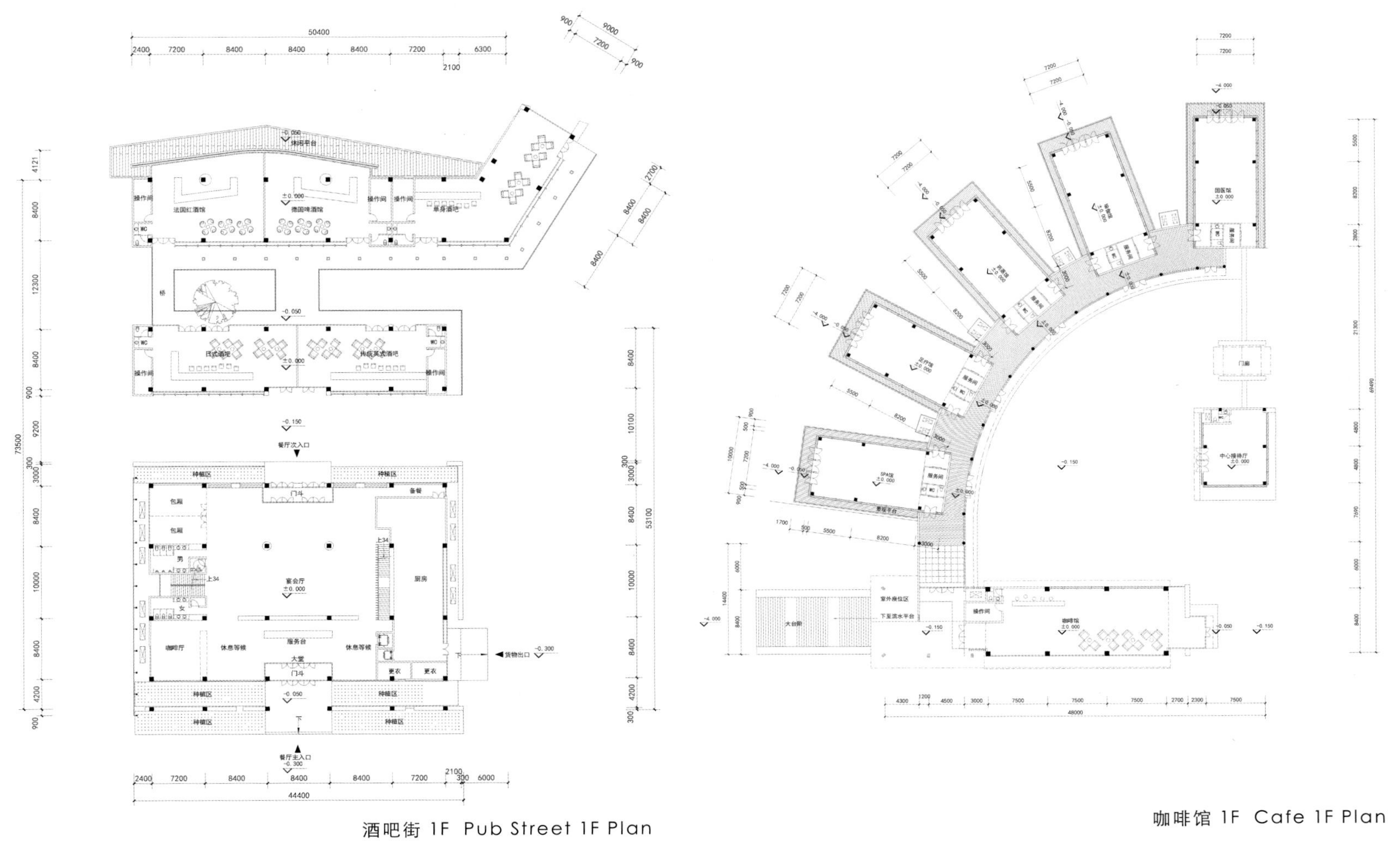

酒吧街 1F Pub Street 1F Plan

咖啡馆 1F Cafe 1F Plan

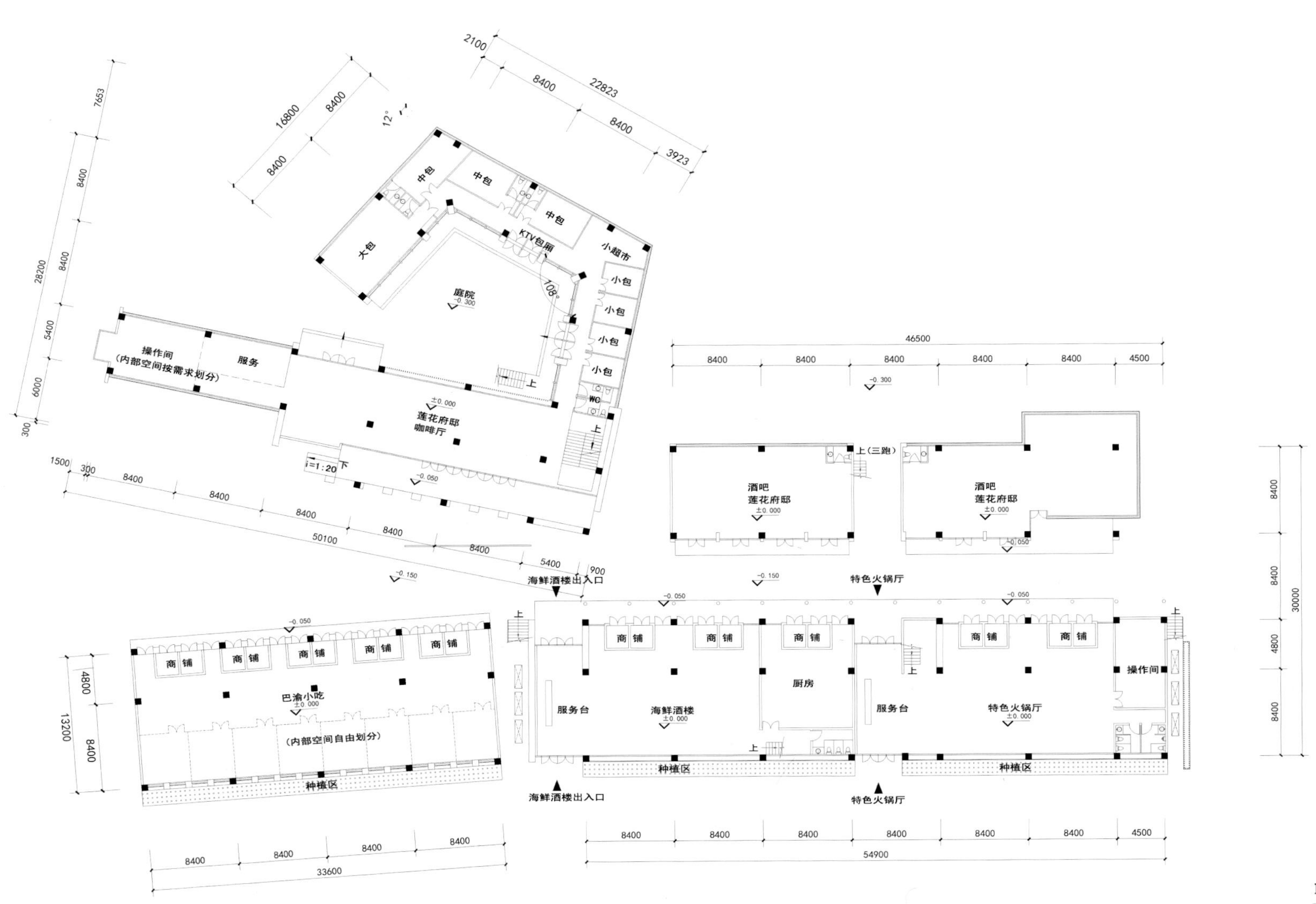

特色火锅 海鲜酒楼 巴渝小吃 1F F&B 1F Plan

建筑设计

风格定位：现代中式风格

设计承袭了中式的特色传统，又融入了现代的建筑理念，把中式的建筑风貌以形神兼备的现代手法充分地表现出来，加之融合了大足独特的文化，形成了独一无二的现代中式商业街风貌。项目建筑风格以现代中式风格为主，以现代的简洁手法结合中式追求内敛、质朴的设计风格，营造现代而又中式的商业空间环境——虚实结合，配以装饰点缀和“留白”处理，体现中国传统的空间精神和韵味。整体环境讲究空间层次感，建筑内部采用虚隔断或者开放式的现代空间手法，给现代中式一种开阔的时尚感；通过现代材料和手法修改传统中的各个元素，并在此基础上进行必要的演化和抽象化，在建筑的整体风格上保留中式的神韵和精髓。色彩以冷灰色为主，配以局部石材，以达到丰富的空间效果。

立面设计：传统元素现代技法

立面采用镂空砖、木色格栅、花格门窗、印花玻璃等丰富的形式和材料表达，局部装饰点缀，局部“留白”处理，繁简有度，给人留下思考的空间，以达到个性鲜明的传统中式与现代完美结合的建筑形象。

Architectural Design

Style Positioning: Modern Chinese Style

The design inherits Chinese characteristic tradition while integrates in modern architectural concepts. Using modern techniques to present Chinese building's appearance in unity of spirit and form. In addition, the Dazu culture is appended in to form a peculiar modern Chinese commercial street. The design utilizes minimal methods to display Chinese restraining and simplicity, created a solid yet void modern Chinese style commercial space. Aligning with leaving blank art technique interprets Chinese traditional spatial spirit and appeal. The whole environment stresses dimensional layering. In interior spaces, virtual partition or open-type methods are used to bring an expansive sense of fashion. Traditional elements are transformed by modern materials and technologies, which is the basis for further evolution and abstraction so as to retain the charm and essence of Chinese style. Cold gray as dominant hue matches sectional stone material to enrich spatial effects.

Facade Design: Traditional Element and Modern Technique

The facade makes use of plenty patterns and various materials such as hollow brick, wooden trellis, lattice door and window and printing glass. Partly decorates and partly leaves blank, keeping a proper complicated and simplified format and leaving enough space for imagination.

小贴士

海棠文化：宋代昌州号称“海棠香国”，大足既是昌州的所属县，又是昌州的治所，所以是“海棠香国”的母地。天下海棠本无香，而昌州海棠却有香，故闻名遐迩，海棠香国也跻身大足八景之一。

石刻文化：大足石刻时间跨度从9世纪到13世纪。这些石刻以其艺术品质极高、题材丰富多变而闻名遐迩，从世俗到宗教，鲜明地反映了中国这一时期的日常社会生活，并充分证明了这一时期佛教、道教和儒家思想的和谐相处局面，是独具特色的世界文化遗产宝库。

Tips

Begonia Culture: Changzhou was named “Begonia Aroma Country” in Song dynasty, and Dazu County belonged to Changzhou and was the Seat of Changzhou State, so Dazu County was the birth-land of “Begonia Aroma Country”. Begonia is without fragrance but aromatic in Changzhou, which makes it enjoy widespread renown. The Begonia Aroma Country is one of eight scenic spots in Dazu County.

Rock Carvings Culture: Dazu Rock Carvings had flourished from 9th century to 13th century. The carving artworks bear high art value and cover myriad themes: common customs, religion and daily social life at that time. And they witnessed a harmonious circumstance of Buddhism, Taoism and Confucianism in our history. The Rock Carvings are unique cultural heritage treasuries in the world.

景观设计：古典意蕴

景观设计意在创造一种传统中国文化与现代时尚元素的邂逅，以内敛沉稳的传统文化为出发点，融入现代设计语言，为现代空间注入凝练、唯美的中国古典情韵。通过对传统文化的认识，将现代元素和传统元素结合在一起，以现代人的审美需求来打造富有传统韵味的景观，让传统艺术在当今社会得到合适体现，让使用者感受到浩瀚无垠的传统文化。使用传统的造园手法、运用中国传统韵味的色彩、中国传统的图案符号、植物空间的营造等来打造具有中国韵味的现代景观空间。在入口处设置入口广场，景观随交通线路自由布置，设置水池、叠水、广场、雕塑、花池、踏步、小品、亭、廊、台、榭等景观小品，并根据景观视线分析，种植不同树木、花草，使之一年四季均有良好的观赏效果，以期达到自然景观与人文景观的最佳效果。

Landscape Design: Classical Connotation

The landscape design creates a perfect blending of traditional culture and modern elements. Traditional landscape gardening, Chinese traditional colors and symbol patterns and planting together build a Chinese modern landscape space. There is a square at the entrance, and along with traveling routes are all kinds of landscapes such as pond, waterfall, square, statue, flower bed, step, landscape sketch, pavilion, porch, stage, shed, etc. Various trees and flowers are planted according to the effect of different landscape visual appreciation, guaranteeing a great viewing all year round, meanwhile achieving a desirable effect of natural landscape and humanistic landscape.

巴渝吊脚楼文化 现代自然主义

重庆缙云山国际温泉度假区

Sichuan and Chongqing Stilted Building Culture　Modern Naturalism

Jinyunshan International Hot Spring Resorts, Chongqing

建设单位：云南心景旅游集团　景观设计：重庆尚源建筑景观设计有限公司
项目地址：重庆市北碚区缙云山十里温泉城　总占地面积：约666 667平方米　一期占地面积：102 359平方米

Architecture Design: Sheenjoy　Landscape Design: A&N
Location: Jinyunshan Tenmiles Hotspring City, Beibei District, Chongqing
Site Area: about 666,667 m²　Phase I Site Area: 102,359 m²

吊脚楼是源于重庆地理及文化等因素而衍生出来，具有重庆特色的建筑，是重庆建筑文化中重要的组成部分。项目依山就势，采用重庆独有的吊脚楼文化和爬坡建筑形式，充分传达了重庆不同层面的文化底蕴和特色。结合现代自然主义建筑风格，充分利用自然资源，使建筑与自然相融合，与中国传统建筑文化中天人合一的思想相呼应。

Stilted building is a crucial component in Chongqing architectural culture, because it generated from local geographic and cultural conditions. The project fully refers to mountainous advantages and stilted building and slope architectural forms to present Chongqing cultural foundations and features. Naturalism architectural style is blended in the design, so natural resources are thoroughly used in the design so as to echo the principle that Chinese traditional architecture focuses a unity of nature and human being.

总平面图 Master Plan

项目背景

重庆市从2009年开始，依托北碚缙云山丰富的自然生态历史人文资源，将山、水、泉、林、地五位一体大手笔规划，打造占地8.99平方公里的北碚十里温泉城。荟萃温泉度假、文化寻根、康体养生、运动休闲、展示交易、滨水游乐、小镇旅游、养生居住等多功能、多形态于一体的国际级休闲度假旅游目的地。

缙云山位于重庆市北碚区境内，坐落在嘉陵江小三峡温塘峡西岸，是国家级自然保护区、国家级风景名胜区和国家AAAA级旅游区，同时也是重庆市植物园所在地。缙云山九峰挺立，拔地而起，山上古木参天，翠竹成林，环境清幽，景色优美，因而有“小峨眉”之称。

缙云山自古以来就是重庆人休闲、娱乐、养生的第一目的地，被誉为重庆的后花园。

规划布局

项目分两期开发，一期由心景缙云•泉渡温泉度假酒店（国际五星级）及高端度假精装公寓、温泉会议中心、艺术运动公园和艺术文化交流中心构成，二期由心景•禅佛禅主题度假酒店（超五星级）、高端别墅社区构成。整个片区按照一心、一带、四片区进行规划布局，形成以心景缙云•泉渡温泉度假中心为核心、运河游乐带为中轴、四大主题片区相互衬托的总体布局。

Project Context

Chongqing has started to plan the Jinyun Mountain rich resources from 2009. The Tenmiles Hotspring City covers 8,990,000 m² and enjoys mountain, river, spring, trees and land of the Jinyun Mountain. It is an international level holiday resort including multifunction of hot spring, literary appreciation, health maintenance, sports, exhibition, water amusement and small town tourist.

The Jinyun Mountain situates in Beibei District, Chongqing, facing the east shore of the Wentang Gorge which is also named Small Three Gorges on the Jialing River. The resort is a National Nature Reserve, a National Scenic Spot, a National 4 A Tourist Area and the Chongqing Botanic Garden. The Jinyun Mountain has 9 peaks soaring with ancient trees and verdant bamboos. Due to its beautiful and tranquil environment, the Jinyun Mountain is also called “small Emei Mountain ”.

The Jinyun Mountain has been the most ideal resort for relaxation, amusement and health maintenance for Chongqing people since ancient times, so it is also called the backyard garden of Chongqing.

Planning Layout

The overall project is divided into two phases. The first phase projects include a Jinyun • Quandu Hot Spring Resort Hotel (international five-star), a top grade refined decoration holiday apartment, a Hot Spring Conference Center, an Artistic Sports Park and an Artistic and Cultural Exchange center; the second phase projects are a Zen and Buddhist Holiday Hotel (super five-star) and a top grand villa community. The whole site consists of one center, one region and four districts that is a hot spring holiday center, a canal amusement region as central axis and four themed districts.

建筑设计

项目采用现代自然主义建筑风格，将重庆独有的吊脚楼、爬坡建筑形式在项目中充分展现。一期产品将重庆独有的吊脚楼风格建筑和坡地景观充分结合，融入温泉养生文化，打造具有重庆地域文化特色的高端温泉度假产品。酒店式温泉汤屋更是由6栋各具巴渝本土建筑特色的江畔吊脚楼组成，每栋都代表了重庆不同层面的文化底蕴和特色。

Architectural Design

The hotel adopts a modern naturalism architectural style, and fully develops Chongqing stilted building form and climbing architectural model. The first phase construction combines stilted buildings with slope landscapes, meanwhile integrates hot spring health culture, creating a regional featured top grade hot spring holiday product. Hotel hot spring resort is composed of 6 Sichuan and Chongqing local characteristic waterfront stilted buildings, and each one stands for a typical kind of cultural foundation and feature.

小贴士

小青瓦、坡屋顶、白灰墙、雕花窗等虽是重庆传统民居的主要建筑元素，但爬坡上坎亦是山城重庆的一大特点，而基于地形而建的建筑，也形成了爬坡式的建筑形式，形成高低错落有致的建筑排布，使建筑很好地融入到自然之中，并具有良好的景观视野。

Tips

Black tile, slope roof, white and gray wall and lattice window are main building elements in Chongqing traditional dwellings, while slope architecture is indeed the first noticeable feature in Chongqing architecture. Buildings are arranged according to landforms in dislocated configuration, so buildings well blend in nature and enjoy good landscape vision.

景观设计

项目以“心净——褪尽浮华、洗尽尘埃，让身心得以释放”为设计理念，将繁杂的思绪塞进简约的空间，通过简洁的设计手法，营造出精致、宁静、休闲、禅意的空间氛围，所有这些空间氛围旨在让人得到彻底的放松，这亦是设计中所追求的最终目标。同时通过运用现代的材质和制作工艺，创造出充满巴渝文化氛围的空间，旨在将巴渝文化通过简洁的方式使空间呈现少一分厚重、多一分闲逸。

项目场地面向运河，背靠缙云山，拥有完美的山体形态。而作为原生山体，设计希望尽可能地减少对山体形态的破坏，规划设计顺应山体肌理，以最大限度地保护山体的原生形态。并设想应如同在原生的山林间，将建筑轻轻地放在景观里，使山体、建筑、景观完美地融合，从而使项目具有重庆山地度假区的特点，并且突出地域特色。无论是场地中为突显空灵感，营造纯净氛围的无边水景，还是藏于林中，使人放松筋骨、享受宁静的幽静汤池；无论是主入口为表现尊贵感、仪式感，而设计的“台”的空间营造，还是公寓间为享受唯美氛围而打造的浪漫花径。每一处每一景都是设计考究、精心打造，呈现出低调奢华、精致内敛的生态环境氛围。

Landscape Design

“Nourish your body and mind. Slow down your life just in sheen joy SPA” is the design conception of the project. It is an exquisite, tranquil, relaxed and Buddhist space that complicated thoughts will vanish in the minimal space, and here people can enjoy an experience of thoroughly relaxation, which is also the ultimate goal of the design. Modern materials and technologies are applied in Sichuan and Chongqing cultural architecture to present more leisure while less heavy.

The project faces a canal and sits against the Jinyun Mountain. As it is an original ecological mountain, the design reduces to transform it as far as possible and complies with mountain texture to present its original appearance to the uttermost. The architecture is like growing in the mountain, harmoniously blending with mountainous landscapes, so the project enjoys the features of Chongqing mountain resort, highlighting regional characteristics. Endless waterscapes bring vanity inspiration; peaceful hot springs hide in forest and create a relaxed and tranquil place; the main entrance presents an exalted feeling and a sense of ritual; space construction is like a “table”; and romantic flower paths are around the apartments, managing an aesthetic atmosphere. Every landscape and every detail elaborate carefully to present a noble, comfortable and elegant ecological environment, featuring in low key, luxury, delicacy and connotation.

岭南画意

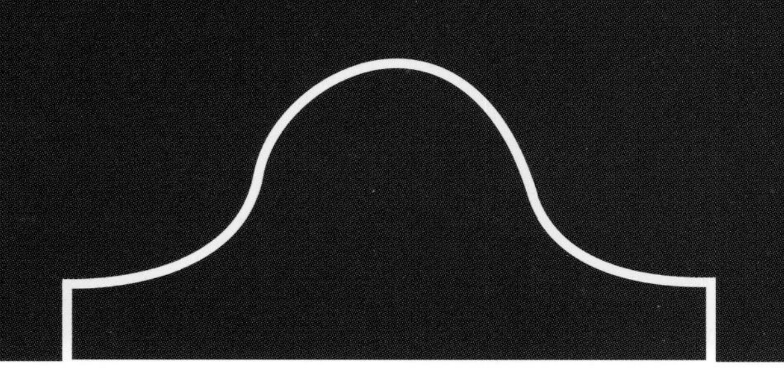

Lingnan Scroll

岭南主要包括广东、广西和海南三地，受地理气候环境、民族风情、文化传统等诸多因素的影响形成了独具特色的岭南建筑文化。因岭南的绝大部分地区位于亚热带沿海区域，所以在建筑材料上以防水材料为主；在功能上具有隔热、遮阳、通风、抗风的特点，同时注意空间的开敞性和流动性。同时，因岭南所处地理位置的特殊性，在建筑文化上呈现出多元性、海洋性和商业性的特点。

在此，我们选择了广东、广西、海南三地的多个项目，从多元性、时代性、民族性及海洋性等几个方面来阐述项目的地域特色以及地域特色中所蕴含的传统性和现代性。

岭南建筑在根源上是江南园林的一部分，亦有着小桥、流水的诗意以及黑白灰的建筑美感。同时又因地制宜，融入当下建筑技术，将传统性与现代性合二为一。

同时，又因岭南地处沿海，而外来文化尤其是下南洋的历史潮流，对岭南的建筑文化也产生了一定的影响，形成了中西合璧的特点。如鹤山十里方圆，虽然呈现出天人合一的传统建筑文化特点，但同时又呈现出东南亚的建筑风格；海南亚泰雨林度假酒店，则是利用山地、森林的优势的同时，采用了东南亚风格，更好地营造出热带雨林度假风情，它们都有着对传统中式建筑文化对天人合一的追求，同时也融入历史因素及外来文化，呈现出当下的建筑特点。

Lingnan region includes Guangdong, Guangxi and Hainan. The region owns unique Lingnan architectural cultures because of geographic climate, national flavor and cultural traditions. Lingnan region is in the subtropic coastal zones, so architectural materials have functions of heat insulation, sunshading, areation and windstanding features; spatial layout concerns openness and flowability. Due to Lingnan region peculiar geographic location, its architecture emerges diversity, oceanity and commercial characteristics.

In this chapter, projects are from Guangdong, Guangxi and Hainan, and we will interpret these projects' regional characteristics from aspects of diversity, times, national characters and oceanity, as well as traditions and modernity being contained in regional characteristics.

Lingnan architecture is a part of Jiangnan garden architecture, and it flaunts bridges and streams poetic feelings and black, white and gray color beauty. Its design acts according to local conditions and introduces modern advanced technologies, combining tradition and modernity as a unity.

The Lingnan region situates in coastal zones, and it has a period of special history that our Chinese ancestors had sailed south to find a way for living, which has influenced Lingnan architecture to get a combination of Chinese and Western elements. Project Guanlan Villa, A Cover of Miles presents traditional architecture cultures yet in Southeast Asian architecture style; Wuzhishan Yatai Rainforest Resort takes advantages of mountainous and forest resources and Southeast Asian style as well to create a tropical rainforest holiday flair. These projects pursue traditional Chinese architectural cultures, while also combine in historic elements and foreign cultures, forming modern architectural features.

岭南民居底蕴 肇庆古朴村落
广东肇庆宋隆小镇文化村
Lingnan Dwelling Foundation　Zhaoqing Pristine Village

Song Long Town Culture Village, Guangdong

开发商：高要市亨昌实业投资有限公司　建筑设计：广州市天作建筑规划设计有限公司
占地面积：2 605平方米　建筑面积：2 340平方米　容积率：0.19　绿化率：35%　采编：谢雪婷

Developer: Heng Chang Industrial Investment Co., Ltd.
Architecture Design: Teamzero Architecture Design & Urban Planning Co., Ltd.
Site Area: 2,605 m²　Building Area: 2,340 m²　Plot Ratio: 0.19　Greening Rate: 35%　Contributing Coordinator: Xie Xueting

基于肇庆作为国家历史文化名城及优秀旅游城市，岭南气息浓郁，自然生态资源丰富，项目在设计上从肇庆地域出发，在建筑设计上以岭南建筑文化为根基，在保证建筑遮阳、通风、隔热等基本功能的同时，采用院落式设计，在保证私密性的同时，公共空间又拉近了人与自然的距离。在立面色彩上，以灰色为主色调，素雅而朴实，并结合街巷设计，营造出深邃、幽远的村落感居住空间。在回归岭南民居传统的同时，亦采用了水泥、玻璃、钢等现代建筑材料，与青石、砖瓦、木料等相结合，突显出古朴而不失现代的气质。

Zhaoqing is a historic and cultural city and a Chinese Outstanding Tourism City, enjoying a rich Lingnan vernacular ambiance and abundant natural and ecological resources. Gathering design elements from Zhaoqing and Lingnan architectural cultures, the project design adopts courtyard layout to guarantee private spaces and certain public spaces, which can shorten distance with nature on the premise of basic functions of sunshade, ventilation and insulation. Gray is a dominant hue on facade, emitting out a pristine and elegant flair. The design applies lane system to form a deep and rural village-like living space. The hotel design respects Lingnan dwelling traditions, and meanwhile some modern materials, such as cement, glass and steel are combined with bluestone, tile and wood, flaunting a pristine yet modern breath.

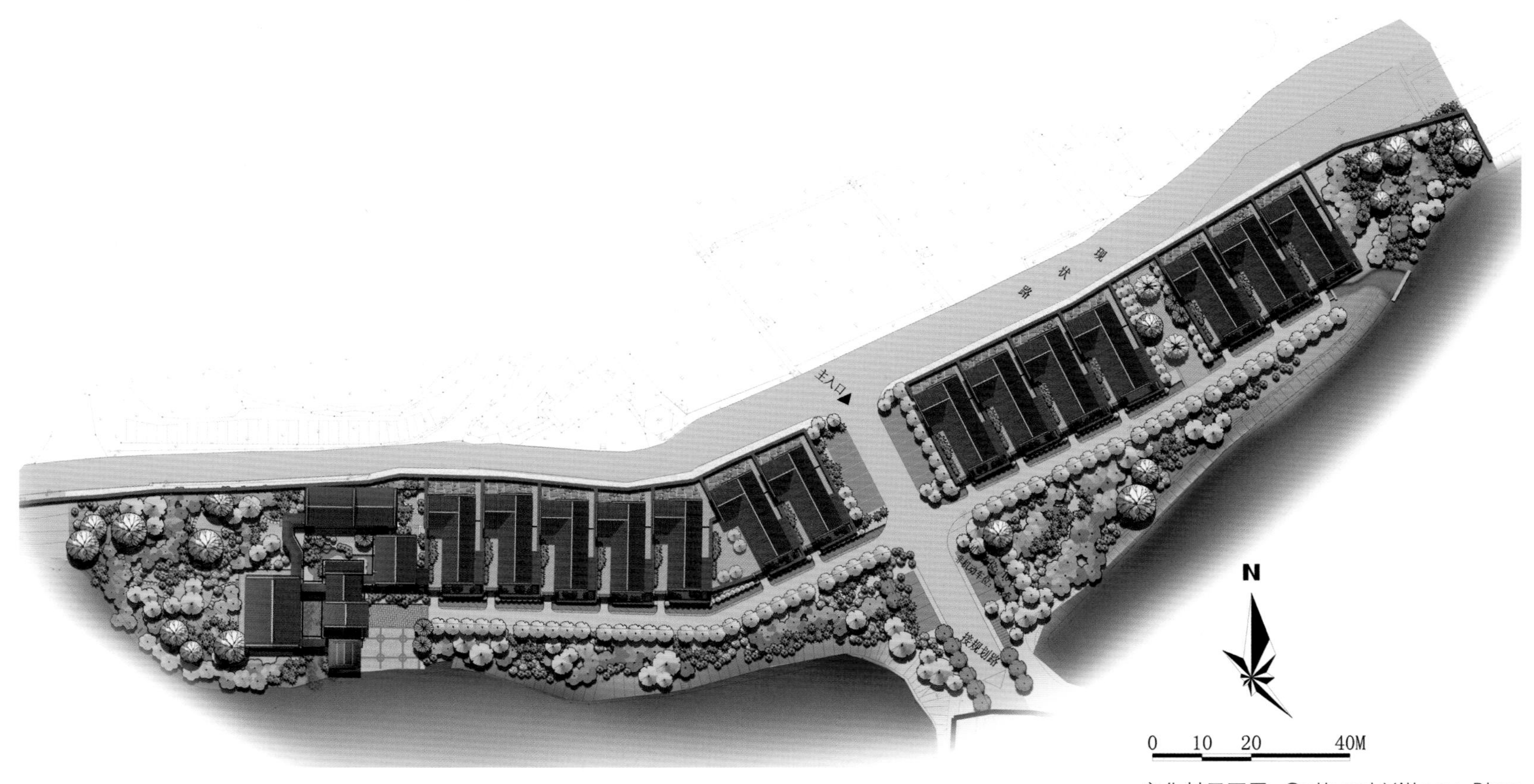

文化村平面图 Cultural Village Plan

创作室平面图 Creation Studios Plan

文化水院平面图 Cultural Water Court Plan

区位分析

项目位于肇庆市高要白土镇金广大道（原广新农业生态园），属于珠三角经济圈范围，交通便利，距离肇庆市区25公里，主要建筑为低层低密度高档文化酒店。

项目背景

肇庆是国家历史文化名城之一，文化底蕴深厚，岭南气息浓郁，是岭南文化、广府文化的发源地和名胜地之一。肇庆也是中国优秀旅游城市、国家园林城市。在建筑文化上，肇庆亦有着岭南建筑文化的底蕴，在功能上具有隔热、遮阳、通风的特点，建筑顶部多采用多层斜坡顶设计，外立面多以深灰色、灰色为主，布局上多设有露台、敞廊、敞厅等开放性空间。宋隆小镇是正域集团重点打造中国首席颐养生态度假岛，总面积约4 500亩的山湖，1 100亩水体，30亩浪漫花海，是集旅游、度假、投资于一体的大型综合地产高端项目。

规划布局：村落形态

项目规划处理上表现了“村”的形态。整个酒店的规划将岭南风格的民居组合成一条村落，通过一条半环形的主道连接起来，各院落内部都有幽静的街巷或小路，院与院之间互相连接，户户有景，以宜人的尺度构成了富有人情味的、古色古香的空间。

Location Overview

The hotel is in Jinguang Avenue (Guangxin Agriculture Ecological Park), Baitu Town, Gaoyao, Zhaoqing. This district belongs to the scope of the Pearl River Delta Economic Circle, so it enjoys convenient transportation, just 25 km away from Zhaoqing downtown. The main buildings in the park are low-rise upscale cultural hotel with low density.

Project Context

Zhaoqing is one of historic and cultural cities in China. It has profound cultural foundations because it is the birthplace and flourishing area of Linnan culture and Cantonese culture. Zhaoqing is recognized as Chinese Outstanding Tourism City and National Garden City. As for architectural culture, Zhaoqing features in Lingnan cultural characteristics that architectural functions highlight heat insulation, sunshade and ventilation; roof is used to adopting multilayered sloping roof with gray facade; and layout applies open spaces, such as terrace, loggia and open hall. Song Long Town is a healthy maintenance ecological resort island constructed by Zheng Yu Group, which covers approximately 3,000,000 m² lake, 730,000 m² waterscape and 20,000 m² flower bed. It is a large-scale integrated project including tourism, holiday and investment.

Planning Layout: Village Form

The planning organizes the park to be a “village”. The overall planning assembles Lingnan style dwellings as a belt of village, connected by a semi-annular main road. Every courtyard has sequestered lane or path connecting to one another. Every house enjoys unique landscapes and all in agreeable scales that constitute an antique and humanistic space.

45m² MINI House 大有世界
45m²山湖精装酒店公寓
3万首付做景区房东
销售中心
400-888-8533

宋隆小鎮
SONG LONG TOWN
大道自然 万象天和
正域

小贴士

冷巷是岭南传统建筑的精髓，在建筑设计中具有组织自然通风的功能。冷巷给封闭的居住环境带来了阳光、空气以及绿体等自然因素，有效地改善了居住空间的舒适环境，也有利于住宅建筑的节能。

Tips

Empty lane is quintessence in Lingnan traditional architecture and bears the function of natural ventilation. Empty lanes inject sunshine, air and greenness into enclosed residential spaces and improve living environments, and it is also in favor of energy conservation.

建筑设计：现代版岭南民居

酒店的院落中设有前庭、中庭、后院，强调岭南民居中院的感觉，院落与院落之间的连接组成大院，提供更大的休闲空间，在岭南闷热的天气里能起到很好的通风效果。

同时，设计利用原来河流地形上的特点，通过主道连接，错落有致地把院落散点在地上，更能体现岭南民居的中式风格。而多条冷巷的设计，起到通风、导风和带走水汽、降低墙体温度的效果，突出“幽”的意境，回应岭南气候的同时，突显村落感。在色彩上，设计采用素雅、朴实的颜色，穿插少许亮色，渲染“素”的意味。在建筑材料的应用上，项目也在传统的民居建筑上做出了现代的改良，使青石、砖瓦、木料的传统材料与水泥、玻璃、钢等现代建筑材料并存，最终塑造出一种古朴、典雅又不失现代气质的建筑。

Architectural Design: Modern Version Lingnan Dwelling

The hotel equips front court, atrium and rear court, which reveals the importance of court in Lingnan dwelling. Courts connect up into compound courtyard where people can enjoy larger leisure area and better ventilation environment in sweltering climate in Lingnan.

The design utilizes original fluvial topography features to dislocate courts while connect by main roads. The layout reflects Chinese architectural style of Lingnan dwelling. Many empty lanes are usefor ventilation, moisture evaporation and lowering temperature of walls. Meanwhile they present secluded feeling, echoing Lingnan climate and a village feeling. As for color application, it adopts plain and elegant colors and a few bright colors to serve as a foil to highlight native flavor. As for architectural material application, it modifies traditional dwelling material combination that bluestone, tile and wood and modern cement, glass and steel coexist harmoniously, creating a pristine, elegant and modern breath.

景观设计

项目园林设计突出步行的氛围，整体景观设计以岭南气候为出发点，没有采用大面积的集中园林和绿化，而是将其打散在周围，化整为零，边边角角都不放过，通过种植众多树木花草，在阳光太强时提供更多的乘凉空间。院子景观设计时避免了大片的绿色草地，而是种植了大量的植物，如竹子、杨树、柏树、肇庆的市花鸡蛋花，组成庭院的效果，并增加假山、亭子，营造富有诗情画意、曲径通幽、峰回路转的公共休闲空间。

Landscape Design

Garden design highlights walking atmosphere. For sake of Lingnan climate, the overall landscape design gives up extensive centralized garden or greening while arranges them in the surrounding areas, breaking up the whole into parts. A large number of trees and flowers are planted at every nook and corner to provide more shade areas under strong sunlight. The courtyard landscapes apply abundant plants instead of grass, such as bamboo, poplar, cypress and plumeria, the city flower of Zhaoqing. These flowers accompanying rookeries and pavilion form courtyards, creating a poetic public leisure space with meandering lanes and winding path.

呼应鹤山发展趋向 打造新东方生态建筑

鹤山十里方圆观澜别墅区

Echo Heshan Developing Trend　Build New Oriental Ecological Architecture

Guanlan Villa, A Cover of Miles, Heshan

发展商：方圆房地产发展有限公司　建筑设计：华阳国际设计集团广州公司　设计团队：郑洲、何狄骏、崔影泉、宋文贤
项目地址：广东鹤山市大雁山风景区　总用地面积：98 181平方米　总建筑面积：34 310平方米　容积率：0.35　采编：谭杰

Developer: Fineland Group　Architecture Design: CAPOL　Design Team: Zheng Zhou, He Dijun, Cui Yingquan, Song Wenxian
Location: Dayan Mountain Scenic Spot, Heshan, Guangdong　Site Area: 98,181 m²　Building Area: 34,310 m²
Plot Ratio: 0.35　Contributing Coordinator: Tan Jie

鹤山市东部城区将建设成为多功能、充满活力的山水园林式的生态城市新区。而项目位于鹤山大雁山风景区西北山麓，享有优越的自然山水资源，因此，项目在设计上亦是充分利用现有自然资源，以新东方度假休闲风定位建筑立面风格，材料上选择暖色调石材、黄锈石、毛石等营造优雅的建筑形体，赋予建筑自然气息。同时，因鹤山地处广东，受东南亚建筑文化影响，在项目设计中亦融合了东南亚建筑风格，并以现代的建筑设计手法，呼应鹤山市打造山水园林式生态新城的意向、融合传统与现代的生态居住空间。

The east area of Heshan will be built to be a new vital multifunctional landscape garden-like ecological city district. The project just locates in the northwest foot of the Dayan Mountain in Heshan, so it enjoys excellent natural landscape resources that will be fully used by the project. It has new oriental leisure style facade, and applies warm color stones, such as palo gold and rubble to build elegant configuration, bringing a natural breath. Heshan is in Guangdong, so the architecture style is influenced by Southeast Asian culture and the project integrates in Southeast Asian style and modern architectural techniques, which echoes the intention of a new landscape garden-like ecological city and a living space combining tradition and modernity.

主要技术经济指标表

地块	项目	数值
B08	规划总用地面积(平方米)	116482.22
	规划净用地面积(平方米)(计算容积率)	96431.10
	总建筑面积（平方米）	51770.67
	计算容积率建筑面积（平方米）	34150.84
	地下室建筑面积(平方米)	17619.83
	总户数	43
	建筑基底面积(平方米)	17784.69
	容积率	0.35
	建筑密度(%)	18.44
	绿地面积(平方米)	47478.42
	绿地率(%)	49.24

总平面图 Master Plan

区位分析

鹤山十里方圆观澜地块作为鹤山十里方圆项目的一部分，位于广东省鹤山市大雁山风景区西北山麓，西北紧邻佛开高速公路，距鹤山市中心区约6公里。周边山体绵延，风景优美。项目基地由东侧大雁山脚向西倾斜，总体地形呈不规则形状，地势起伏，为丘陵地貌，高差约20～40米。地形高程起伏变化较大，部分山体坡度超过30度。

项目背景

鹤山城区是全市的行政、经济、文化中心，珠江三角洲西部重要的水陆交通要冲之一，为工商业发达、环境优美的现代化滨江城市。鹤山市东部城区将建设成为多功能、充满活力的山水园林式的生态城市新区。

项目概况

项目是以度假、休闲为主导的高端别墅区，由临水A型别墅、观水B型别墅、山景N型别墅、园景P型别墅组成，共43席。项目地块与大雁山景区相连，有山水直接流入地块，在地块内形成多处低洼的水塘，具有良好的储水条件。

Location Overview

Guanlan Fineland plot is a part of Guanlan Villa project, and it situates in the northwest of mountain foot of the Dayan Mountain Scenic Spot, Heshan, Guangdong. It is close to the Foshan-Kaiping Expressway, about 6 km away from Heshan downtown. It is surrounded by continuous mountains, enjoying beautiful sceneries. The project site descends from the east of the foot of Dayan Mountain to the west in irregular shape. The hilly terrain is undulating with altitude difference of 20 to 40 m, and parts of mountain gradient are over 30°.

Project Context

Heshan urban area is the center of administration, economy and culture of the city. Heshan is one of an important amphibious transportation pivots with developed commerce and a beautiful modern riverside city. In the east of Heshan, a multifunctional and vigorous garden-like ecological district is to be developed.

Project Overview

The project is an upscale villa area for holiday and relaxation, including Waterfront Villa Model A, Water-view Villa Model B, Mountain Villa Model N and Garden Villa Model P, amount to 43 suites. Because the site connects to the Dayan Mountain Scenic Spot, the mountain water naturally flows into the plot and forms many low-lying pools with good water-storage capacity.

设计理念

在方案设计中，提出“现代又自然”的设计理念，试图用简约现代的建筑手法、质朴自然的材料语言，将传统的东方建筑与度假风情相融合，创造出全新的起居空间体验——还原大自然风情、强调简朴归真，营造舒适亲切的度假氛围。

规划设计

项目用地地形为北高南低的缓坡，三面环湖，用地条件优异。规划中的难点是项目容积率虽不高，但户型的基底面积较大，栋数较多，容易造成住宅过于密集、公共空间过于零碎的问题。项目尊重地形，用扇形的曲线布局沿湖面伸展，每两户间留出较大的东西间距使得南北视线通透；双边服务的道路大大减少了交通面积，边角用地化零为整，转变为集中的景观用地——北侧绿化隔离带、东侧主入口景观廊，南侧共享水院，大大提升了公共景观品质。

Design Conception

In the process of the planning, the designers propose a concept of “Modern but Nature” and strive to merge holiday ambiance into traditional oriental architecture with simple but modern architectural skills and plain but natural materials, thus the design creates a brand-new living space so as to restore the beauty of nature, highlight returning to original simplicity and form a comfortable and cordial resort atmosphere.

Planning Design

The project plot is a gentle slope descending from north to south, with lake at three sides. The planning difficulty lies in low plot ratio and large area of house type base, so it exists too many buildings to share spacious space and public area. Under this condition, the project design still respects the existing terrain and arranges buildings along the lake in a sector layout, meanwhile, it leaves large east-west orientation spaces between every two buildings, thus south-north orientation vision is transparent. Besides, bilateral service roads largely reduce traffic area, and corners around them are transformed into concentrated landscape areas: greenbelt in the north, water courtyard in the south and landscape corridor at the east main entrance. All these designs greatly level up the quality of public landscapes.

立面设计

项目采用以新东方休闲风为主调的立面造型风格，通过现代建筑设计手法，融合传统东南亚建筑风格，营造出大自然风情，强调简朴、舒适亲切的度假氛围。

为满足高端别墅的定位需求，在建筑材料选用上进行升级。墙身以暖色调石材为主，主体采用黄锈石来打造主体色调，通过不同的组合形式、实与虚的对比，营造大气优雅的建筑形体；通过毛石围墙的应用，体现新东方休闲风格还原自然的设计原则，使建筑更好地与环境相结合，塑造简朴、舒适的度假风情。上部采用铝合金木纹板和木百叶，整体风格通透清爽；基座采用毛石等天然石材，以粗犷的质感与主体形成轻与重的对比。

此外，项目沿用度假建筑平缓深远的大屋顶形式，以轻盈的屋顶与厚重的体块形成对比。通过阳台和露台的巧妙结合，丰富建筑内部空间的连续性和多样性，为景观环境的塑造提供平台，进一步提高居住生活品质。室外庭院与观景水池、泳池进行结合设计，营造良好的度假氛围，进一步体现新东方休闲风格更人性、亲切和奢华的品质感受。

Facade Design

The facade features in new oriental leisure style. The villa areas appear natural, simple and comfortable holiday atmosphere due to modern architectural technologies combining traditional Southeast Asian architectural style.

To meet the demand of upscale villas, the designers upgrade the construction materials. The wall mainly applies warm-colored stone material – palo gold. Grand and elegant architectural body turns out by various combining forms and void and solid contrast. The rubble wall reveals a new oriental leisure style that complies to natural design principle, and meanwhile it makes buildings integrate well into the environment, creating simple yet comfortable holiday flair. The upper part of the facade employs aluminum-alloy-wood board and wooden blinds to bring a refreshing feeling. Natural stone rubble is used in foundation bed, and its rough texture forms a contrast of lightness and heaviness to the main body.

Furthermore, the project follows a large gentle-slope roof commonly used in resort building, thus forms a distinctive contrast of light roof to heavy volume. Balcony and terrace are combined smartly to enrich continuous and various interior space and provide a platform for landscape design so as to level up life quality. Outdoor courtyard, viewing pond and swimming pool are also combined together to create favorable holiday atmosphere, and meanwhile further reveal a cordial, luxurious and humanized quality feeling of new oriental leisure style.

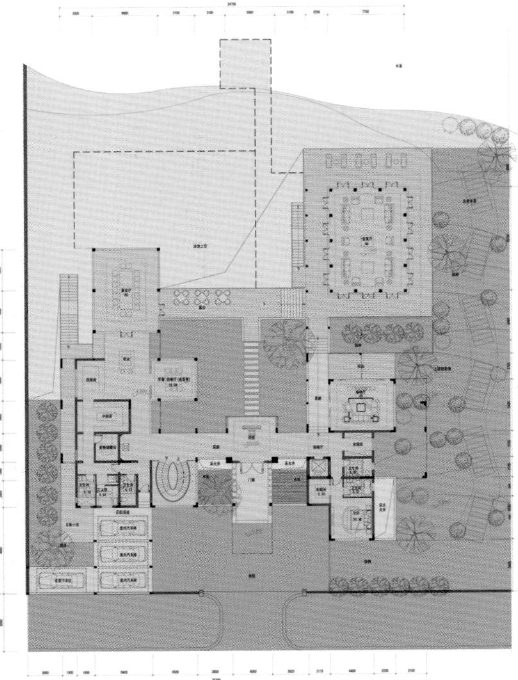

A型别墅一层平面图 Villa A 1F Plan

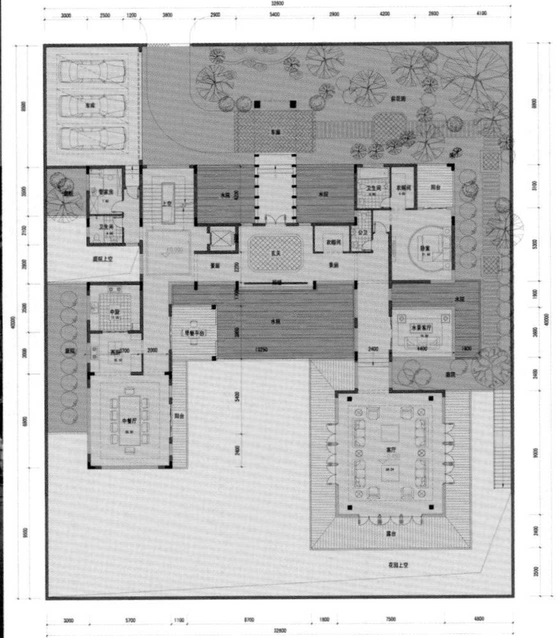

B型别墅一层平面图 Villa B 1F Plan

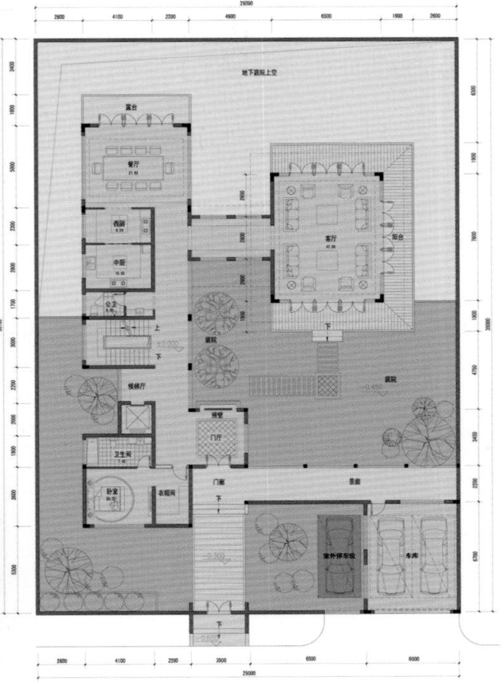

P型别墅一层平面图 Villa P 1F Plan

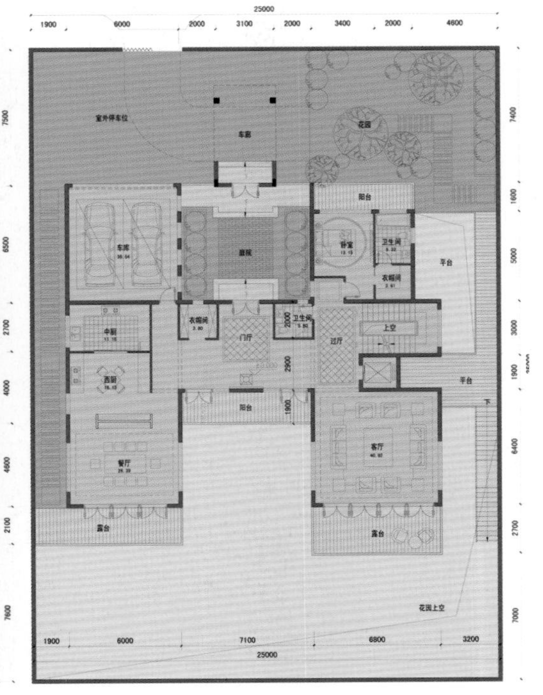

N型别墅一层平面图 Villa N 1F Plan

户型设计

户型面积为600～1 000平方米。为避免大户型常见的体量集中和尺度过大的问题，设计围绕核心庭院在场地上舒展铺开平面，以景廊、院落连接起居空间及主要功能房间，在动线设计上考虑移步换景和对景，使得各个功能房间有敞开、半开放或私密的不同景观体验。

此外，各户型体块简洁，面宽较大，为着重突出入口位置的尊贵感，设计中以铝合金木纹百叶构件为元素，在建筑二层形成连贯的体量和节奏，以局部的繁密装饰与整体立面的素净形成繁与简的对比。

House Type Design

Apartment areas range from 600 m² to 1,000 m². To avoid common problems like compact volumes and over-sized space in large-sized house, the house type design focuses on stretching out central courtyard and using landscape corridor or courtyard to connect up living spaces and main functional rooms. Circulation design strives on obtaining "one step one view" and view-responding effects, which brings all sorts of functional rooms with various landscape experience of open, semi-open or private landscapes.

In addition, with the existing conditions of simple volume and large area, every house type employs aluminum-alloy-wooden blinds to highlight a magnificent entrance. On second floor, facade design adopts continuous volume and rhythm, complementing with dense adornments on parts, which forms a contrast of simplicity and complication to the overall facade.

热带雨林主题酒店 五指山黎苗特色建筑

海南五指山亚泰雨林度假酒店

Tropical Rainforest Theme Hotel　Wuzhi Mountain Ethnic Characteristics

Wuzhishan Yatai Rainforest Resort, Hainan

开发商：亚泰集团　建筑设计：深圳市建筑设计研究总院有限公司　景观设计：深圳夏岳环境艺术设计有限公司　室内装饰：上海迅美装饰设计有限公司
大堂装饰设计：广州利杰登酒店设计公司　项目地址：海南省五指山市水满乡五指山国家级自然保护区内
占地面积：95 614平方米　建筑面积：24 230平方米　容积量：0.25　绿化量：75.5%　采编：谭杰

Developer: Yatai Group　Architecture Design: Shenzhen General Institute of Architectural Design and Research Co., Ltd.
Landscape Design: Shenzhen Shine Design Co., Ltd.　Interior Design: S.A.M　Lobby Decoration: Guangzhou Legend Hotel Design
Location: Shuiman Village, Wuzhishan City, Hainan　Site Area: 95,614 m²　Building Area: 24,230 m²
Plot Ratio: 0.25　Greening Rate: 75.5%　Contributing Coordinator: Tan Jie

五指山市位于海南岛中南部腹地，是黎族、苗族聚居地，具有浓郁的少数民族气息，其中五指山是海南第一高峰。五指山亚泰雨林度假酒店位于五指山上，享有丰富的自然景观资源，并深受少数民族文化的影响，在建筑上不仅充分利用自然景观资源，亦将黎苗族建筑文化——船形屋与吊脚楼融入其中，采用天然材料与自然色调，结合暖黄色的Stucco外墙涂料与高级文化石基座以及木制或仿木质材料，实现建筑、自然的天人合一的境界，并设计大坡屋顶突显东南亚热带雨林的度假风格。

Wuzhishan City situates at the hinterland of south central of Hainan where Li and Miao ethnic minorities have settled from ancient, so it is rich in ethnic flavors. The Wuzhishan Mountain is the highest mountain in Hainan, and Wuzhishan Yatai Rainforest Resort occupies at the foot of the grand mountain, enjoying opulent natural landscape resources. The architectural design is affected by ethnic culture, because it not only takes full advantage of natural resources, but also takes in Li and Miao ethnic minorities elements: boat shape house and stilted building. Natural materials combine with natural colors. Warm yellow Stucco exterior wall coating matches advanced cultural plinth with wooden or faux-wooden materials, realizing a unity of architecture, nature and human being. Grand slope roof brings a style of Southeast Asian rainforest resort.

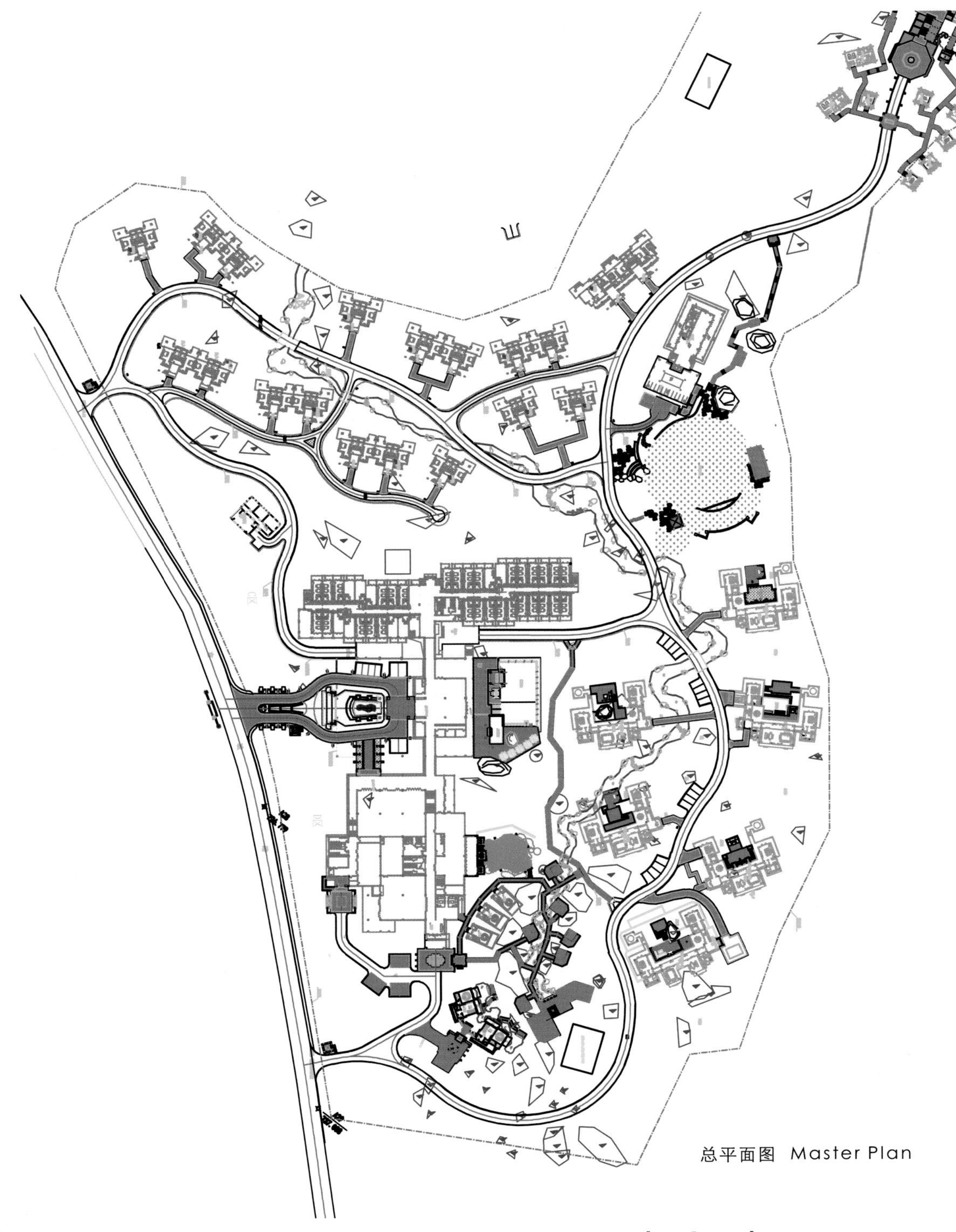

总平面图 Master Plan

区位分析

五指山亚泰雨林度假酒店坐落在海拔1 867米的海南第一高峰——五指山山脚下，酒店周边环绕着原始茂密的热带雨林，雨林中有沟谷小溪，有激流飞瀑，空气清新甜润，负氧离子平均含量达到30 000个/立方厘米以上，是一处养生休闲度假的好去处。

项目背景

五指山市位于海南岛中南部腹地，是海南岛中部地区的中心城市和交通枢纽，也是海南省中部少数民族的聚居地，黎族和苗族便位列其中。黎族传统居住的房屋以茅草为建筑材料，房屋骨架用竹木构成，黎族茅草屋主要有船形屋和金字形屋。一般利用大自然的原材料，如茅草、木料、竹子、红白藤、山麻等。木料则多用格木。船底形房屋又有铺地形和高架形之分，房门开在房屋的两端。而苗族家人喜欢木质建筑，因地势原因，以苗家吊脚楼建筑尤为独特。

Location Overview

Wuzhishan Yatai Rainforest Resort locates at the foot of the Wuzhishan Mountain-the highest mountain in Hainan with an elevation of 1,867 m. The hotel is surrounded by thick primitive tropical rainforest where brooks, ravines and plunging waterfalls bring fresh air and negative oxygen ion average content of 30,000 per cm^3 , so the hotel is an ideal place for a holiday.

Project Context

Wuzhishan City situates at the hinterland of south central of Hainan, and it is a central city and transportation hub in the midland of Hainan. It is inhabited by many ethnic minorities, such as Li and Miao nationalities. Li nationality traditional dwelling is thatch building that is constructed by thatch and wood in boat pattern or glyph of gold. Construction materials are from nature: thatch, wood, bamboo, red and white cane and alchornea. Wood materials mainly choose erythrophloeum ferdii. Boat bottom shape dwelling has two patterns, and both of them have doors at each side of a house. Miao nationality likes wooden house, and due to special terrain, stilted building is typical architecture at Miao ethnic minority settlements.

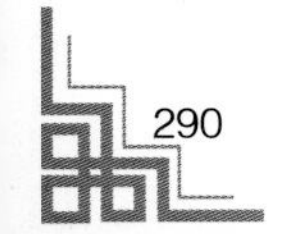

设计理念

项目为旅游、度假、休闲、养生等多功能的山地高星级酒店，充分利用山地和森林的天然优势，结合地形地貌，提供一种通透的、可向户外拓展的建筑空间，打造一个充满热带雨林风情，同时符合旅游发展和康体养生需求、展示绿色原生态的度假酒店。并采用司法自然的空间布局理念，注重景观与建筑群体之间的相互关系，尽可能提供更多的休闲环境空间，形成一系列尺度合理、富有韵味、掩映在丛林之中独具特色的酒店建筑群体。

规划布局

酒店主楼位于项目基地中部，中间为酒店大堂，左右两侧分为C区四层的标准间客房区和D区两层餐饮会议功能区（含户外5栋VIP中餐包厢）；A区为12栋两层楼别墅客房区；B区为6栋一层式豪华别墅客房区；E区为1栋一层休闲吧及书吧；F区为11栋养生SPA区。

酒店配有中餐厅、西餐厅、精品餐厅、多功能厅、会议室，康体设施有健身房、台球室、壁球室、棋牌室，户外有游泳池（含更衣室）、VIP包厢、雨林吧、儿童游乐场。

Design Conception

The project is a star level hotel providing traveling, holiday, relaxation and health maintenance functions. It takes advantages of mountain and forest here, meanwhile complies with terrain to build a transparent and outward development space. Tropical rainforest style brings a primitive green ecological resort hotel. Natural spatial layout guarantees an interaction between landscapes and architecture groups. Recreational environment spaces are sufficient due to proper scale organization. It is a group of unique hotel buildings in forest.

Planning Layout

The main building of the hotel sits in the center of the site. The lobby hall is in the middle; district C has four-floor standard guest room; district D has two-floor F&B and conference functional zones (including 5 buildings containing Chinese meal VIP room). District C and D stand on each side of the lobby. District A consists of 12 buildings of two-floor villa guest room; district B has 6 buildings of one-floor deluxe villa guest room; district E has one-floor amusement bar and book bar; district F has 11 buildings of health maintenance SPA. The hotel has plenty of holiday functional facilities: Chinese meal restaurant, Western meal restaurant, boutique restaurant, multifunctional hall, conference room, fitness center, billiard parlor, squash court, chess and card room, outdoor swimming pool (with changing room), VIP room, rainforest bar, children's playground.

建筑设计

建筑设计最大限度地将建筑与自然景观融为一体，针对亚泰雨林酒店所处的热带雨林气候高温高湿的特点，结合五指山当地气候特征、自然环境以及民族特色，建筑风格偏向于大坡屋顶式的东南亚热带雨林的度假风格。大挑檐适合当地雨水和日照环境的要求，暖黄色的Stucco外墙涂料与高级文化石基座相结合，栏杆和檐口为木制或仿木质材料，建筑局部加入了具有当地黎苗民族特色的符号元素。大量的折叠门让外部景观一览无遗，与内部空间相互渗透。同时，采用能自由呼吸的装饰砂浆来装饰这外立面，并在较大的区域内采用西蒙木纹漆来强化五指山区特色建筑风格——船形层与吊脚楼的建筑元素，让游客在体验野外的热带雨林风光中，置身于奢华与质朴相结合的高端建筑氛围之中，感受天人合一的境界与当地特色的住宅文化。

Architectural Design

The architectural design integrates the buildings with natural landscapes to the extreme. According to tropical rainforest climate features of high temperature and high humidity, natural environment and local ethnic minorities characteristics, the architectural design applies grand slope roof in Southeast Asian rainforest holiday style. Large cornice complies with local raining and sunshine conditions. Warm yellow Stucco exterior wall coating combines with advanced cultural stone plinth. Railings and cornices are made of wooden or faux-wooden materials. Li and Miao ethnic minorities architectural elements are added in parts of the building. Folding doors are commonly used to bring outdoor landscapes in. Decoration Stucco wraps up facade, and Simon wood grain paint expresses Wuzhishan featured architectural elements: boat shape house and stilted building. Tourist can experience a tropical rainforest viewing and enjoy a top-level architectural ambiance integrating luxury and pristine in the same time. In addition, they can feel the harmonious unity of nature and human being and local featured residential culture.

小贴士

STUCCO：其表面平滑，但具有明暗深浅纹理，可调制任何颜色或金属颜色，能营造具特殊肌理效果的艺术墙面，表现出良好的立体釉面效果和层次感。可令墙面光滑柔顺，达到如鹅绒般轻柔光滑的触感。施工随意，漆膜坚硬，防水，具有其他涂料无法比拟的装饰和使用性能。

Tips

STUCCO has smooth surface with light or dark grains in various depth, meanwhile it also can be inserted into any color including golden color to create a wall with special artistic effects that is perfect in three-dimensional glazing and layering, which makes a slippery and gentle wall with a velvet touch. The coating needs no complicated techniques but is good at hardness and waterproof that are beyond comparison over other coatings in decorativeness and operational performance.

桂北民居村落 现代奢华气息
阳朔悦榕庄
North Guangxi Dwelling Village Modern Luxury Breath

Banyan Tree, Yangshuo

酒店开发商：悦榕控股有限公司 建筑/室内/景观设计：Architrave Design and Planning 酒店地址：广西壮族自治区阳朔县福利镇
占地面积：约70 667平方米 总建筑面积：25 000平方米 采编：陈惠慧

Developer: Banyan Tree Holdings Limited Architecture/Interior/Landscape Design: Architrave Design and Planning
Location: Fuli Town, Yangshuo County, Guangxi Zhuang Autonomous Region
Site Area: about 70,667 m² Building Area: 25,000 m² Contributing Coordinator: Chen Huihui

漓江风光、喀斯特地貌是广西独具特色的自然美景，也是不可错过的旅游景点。阳朔悦榕庄将漓江与喀斯特地貌纳入自己的景观视线，通过阳台等设计，将其拉入室内，营造出与自然和谐的空间。而桂北民居错落的独特韵味，融汇现代奢华与当地文化的设计，营造出了一份阳朔独有的度假气息。

The Lijiang River enjoys beautiful landscapes and Karst landform is unique in Guangxi, so both of they are can’t-miss scenic spots. The Banyan Tree hotel design abstracts landscapes of the Lijiang River and Karst landform, and pulls them indoor through balconies so as to create a natural and harmonious space. Dislocated North Guangxi dwelling has its typical flair, and the architectural design combines modern luxury and local cultures to build a unique holiday breath.

LEGEND 图例

Facilities 公共设施

1 Lobby 大堂
2 Banyan Tree Gallery 悦榕阁
3 Yoga Room 瑜伽房
4 Elevator 电梯
5 Business Centre 商务中心
6 Meeting Rooms 会议室
7 Banquet Hall 宴会厅
8 Gym 健身中心
9 Banyan Tree Spa 悦榕SPA

Suites & Villas 套房和别墅

Gardenview Suite/Mountainview Suite 园景套房/山景套房
Garden Villa 花园别墅
Riverside Villa 水岸别墅

Restaurant 餐饮区

1 Ming Yue 明月餐厅
2 Qing Feng 清风
3 Bai Yun 白云餐厅
4 Aqua Cafe 碧水

总平面图 Master Plan

区位分析

阳朔悦榕庄坐落于阳朔县漓江下游福利镇地界，漓江“钻石水道”之畔，距离桂林国际机场大约1小时30分钟车程，距离阳朔县城和码头大约30分钟车程。

项目背景

福利镇历史悠久，且有“三山环古镇，一水抱绿洲”之称。和阳朔县城以及其他古镇不同，福利镇不仅是风光秀丽的旅游名镇，还是全国最有名的国画镇之一，被称为“中国画扇第一镇”。相对于百里漓江，福利镇的江面显得更为开阔，给游人一种天地悠悠然的感觉。百里漓江有九马画山，福利镇则有九马归巢，有着归隐的韵味。九马画山只可远观，九马归巢却可近赏，从小路通往半山腰的岩洞，上面还保留着明朝嘉靖年间的石刻。500多年历史的留公村是福利镇段漓江的又一秘境。村子始建于1487年，依山傍水，山清水秀，至今还保留着明代和清代的建筑，北伐战争时，孙中山先生溯江而上，也曾泊舟留公村。

规划布局

项目由1~3层的建筑围成庭院，为单体别墅型建筑，拥有148间客房。此外，项目还设有大堂吧、泳池吧、餐厅、图书馆、宴会厅等相关配套设施。

Location Overview

The hotel locates in the Lijiang River downstream Fuli Town, Yangshuo county, and it stands by the Lijiang River that is also called a “diamond watercourse”. It takes 90 min drive to the Guilin International Airport and about 30 min to Yangshuo county center and wharf.

Project Context

The Fuli Town is renowned for a long history. The ancient town is surrounded by mountains at three sides and a river flows in front of it, so there a poem says “three mountains encircle an ancient town and a river embraces an oasis”. The Fuli Town is a tourist town with beautiful sceneries, moreover, it is one of the most famous traditional Chinese painting towns in China, and meanwhile it is called “the first town of Chinese painting fan”, which makes it different from other towns in Yangshuo County. Along with the hundred miles Lijiang River, the water surface in Fuli Town is relatively wide to produce a feeling that the world will be lasting forever. The Lijiang River has a scene like nine horses running while the scene just can be seen from afar, however the Fuli Town enjoys a nine horses homing scene, implicating hermitage, and the scene can be experienced nearby. Many stone carvings in the years of Jiajing period of the Ming Dynasty are conserved in the grottoes on halfway of hills. In the Fuli Town, there is a Liugong Village has more than 500 years of history, and it is a paradise by the Lijiang River. The village was built in 1487, however buildings of Ming and Qing dynasties are still conserved well here. In the period of the Northern Expedition, Sun Yat-sen anchored his boat by the village.

Planning Layout

The hotel is in courtyard layout enclosed by buildings of one to three floors. It has 148 guest rooms, and they are in monomer villas. The hotel also provides related supporting facilities: lobby bar, swimming pool, dining hall, library and banquet hall.

建筑设计

建筑设计以传统的桂北民居风格建筑为基调，借鉴了桂北民居小青瓦、坡屋面、白粉墙、木格窗、吊阳台、青石墙裙等经典元素，色彩淡雅朴实，层叠起落的白色马头墙穿插在灰色破瓦之间，配以天然石材基座，高墙窄巷，饰以砖雕，雕花垂柱的门楼鳞次栉比，突显传统村落神韵。

Architectural Design

The architectural design bases on traditional north Guangxi dwelling and abstracts local classic elements, such as Chinese-style tile, slope roof, whitewash wall, lattice window, hanging balcony and bluestone wainscot. The design adopts light colors. Layered white wharf walls insert in light gray tiles on natural stone plinths. Narrow lanes lie between high walls, and they are decorated with stone carvings. Gatehouses with carved vertical columns stand side by side, highlighting a traditional village flair.

室内设计

室内设计融入中国传统建筑设计元素，设计质朴而优雅，142间套房和别墅都充满当地建筑特色。房间室内设计采用黑木、竹子，并选用大理石镶嵌工艺，配备大地色系，用现代手法演绎出中国式的传统典雅；每个房间装点着格子窗和水墨画，摆放了各种手工艺品和中式经典家居，为宾客打造一处真实的心静轩。

此外，通过阳台、落地窗等细部设计，无论是套房还是别墅均能欣赏到自然美景。在园景套房内，宾客能够饱览久负盛名的阳朔喀斯特山脉全景，踏上宽敞的阳台，宾客可近看度假村内的蜿蜒河流，远眺绿色的雄伟山脉；山景套房拥有宽敞无比的室内空间以及阳台，葱郁的喀斯特山峰与清澈的漓江之水相互辉映，美景一览无遗；单卧房花园别墅专为情侣设计，配备宽敞起居室、优雅卧室、日光休闲甲板以及竹林花园，是一处充满浪漫气息的私密之所；单卧房水岸别墅毗邻静谧的漓江，舒适的室内空间被壮美景色萦绕，处处散发着浪漫情调；双卧房水岸别墅专为家庭及好友居住所设计，配备起居室、主卧房、客房、日光休闲甲板、竹林花园，宾客不但可以在别墅周围的休闲区域放松身心，更可欣赏云雾缭绕的山脉以及缓缓流淌的漓江之水。

Interior Design

Traditional architectural elements are introduced in the interior design. The overall project is pristine and elegant, and 142 suits and villas are full of local characteristics. Blackwood and bamboo are local natural materials, and marble inlay technology matches earth tone to bring a traditional elegance; every room decorates with lattice windows, ink and wash paintings, handicrafts and Chinese-style furniture to provide a real tranquil space.

Pushing floor-to-ceiling curtains aside and standing on balcony, guests can enjoy Karst mountains and winding rivers in Gardenview Suite; Mountainview Suite has spacious balcony and interior spaces. Single Bedroom Garden Villa is for lovers and couples, equipping with a large living room, an elegant bedroom, a sunlight deck and a bamboo garden, so it is a private and romantic space; Single Bedroom Riverside Villa is close to the Lijiang River; Double Bedroom Riverside Villa is for family or a group of friends, and it equips with a living room, master bedrooms, guest rooms, a sunlight deck and a bamboo garden so that guests can have fun around the villa and experience scenes of mountains in mist and winding flowing rivers.

滇藏如画

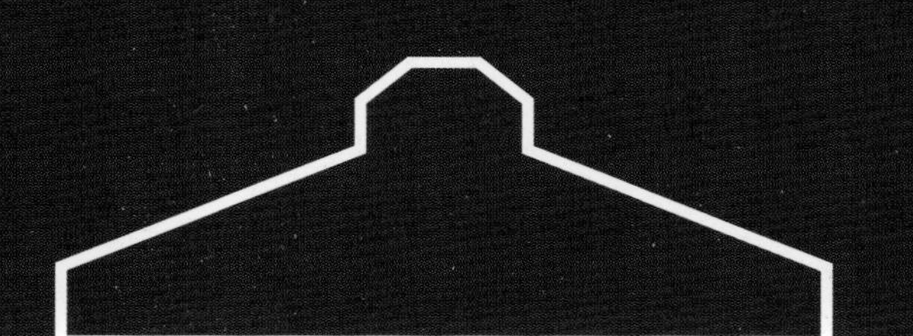

Picturesque Yunnan-Tibet

滇藏地区主要是指云南和西藏地区，这是两个富有原生态气息地区，是很多旅游者的梦想之地。这里有着浓郁的自然气息，给人远离城市喧嚣、亲近的自然之感。在这里，可以忘却尘世烦扰，放下繁重的负担，放飞心情。滇藏的美是自然的、惬意的、质朴的，让人领略到返璞归真的生活。滇藏的建筑，亦如滇藏所带来的自然之感，不仅富有浓郁的自然气息，还带有浓重的民族色彩。

滇藏，是少数民族聚居区，有着深厚的民族文化根基，每个少数民族都有其自己的民族发展历史和民族文化。而对于世居云南的25个少数民族，云南省民语委首先对其民族语言文字进行了整理和归类，使其更为系统化，也更有利于文化的传承和发扬，这也有利于记录和传承云南地区的建筑文化。云南不仅文化呈现多元性，建筑形式亦具有多元性。而西藏地区不仅受少数民族文化的影响，亦受佛教文化影响，佛寺是西藏地区建筑文化的重要组成部分。西藏的传统民居亦是多姿多彩的，有可随处迁居的帐篷，有固定的土木结构的房屋，亦有以石为材的碉房，还有竹楼、木楼和窑洞，其中尤以帐篷和平顶碉房最为常见。可见，在大的环境影响下，滇藏地区的建筑具有原生态之感，在细微之处，不乏民族性。

在此，我们从滇藏地区的自然生态性与民族性为出发点，结合项目的具体剖析，来了解当地建筑的地域性、民族性与多元性。

大理沧海一墅国际养生度假小镇背靠苍山，面向洱海，地势西高东低，其间散落着台阶状的梯田。为了更好地打造原生态度假风情，设计从梯田建筑和民居村落中寻求灵感，将用地划分成若干块“梯田”建筑围合而成的组团式的“村落”，并以风景再生的方式，还原生态景观，与大理常见的景观——村落、梯田相呼应。

丽江金茂君悦酒店设计则以纳西民族文化为根基，将丽江秀美的雪山风光与纳西民族风情融汇于项目之中，让旅客充分体验到自然与民族特色。

昆明洲际酒店在设计上，以现代美学的视角，通过“蝴蝶”主题，展示出项目的生态、自然、优雅之美，并将七彩云南的活泼与绚烂传达给旅客。

云南腾冲悦椿温泉村充分展示了云南乡村风光，在设计上，从基地现状出发，对原有村落以现代美学角度进行修饰、恢复和美化，演绎出富有特色的云南文化现代村落度假空间。

This chapter is about Yunnan and Tibet regions that are many tourists' dreamland, full of original flavors. Arriving here, you will feel a rich natural breath, far away from noisy city yet close to nature, so you can discard mortal bother and heavy burdens and lighten up your mood. The beauty here features in nature, cosey and primitive, and it brings you back to a real life. Yunnan and Tibetan architecture also owns natural feelings and rich ethnic characteristics.

Yunnan and Tibet have many ethnic minority settlements and bear profound cultural foundations. Each minority possesses its own history and ethnic culture. There are 25 minorities localizing in Yunnan, and the Yunnan National Language Committee has sorted out and classified their languages to make them more systematic and be conducive to inherit and promote cultures. In the process of language gathering, Yunnan architecture has been recorded as well. The local architecture appears multiple forms due to its multiculture. The Tibetan region is influenced by minority culture and Buddhist culture, so Buddhist temples are an important part in Tibetan architecture. The Tibetan dwellings have all sorts of forms, such as movable tabernacle, settled civil architecture, stone blockhouse, bamboo house, wooden house and cave dwelling. Among them, tabernacle and flat roof blockhouse are common dwellings. Influenced by macro environment, the Yunnan-Tibet architecture brings us an original ecological feelings, while presents ethnic characters in details.

In this chapter, we would like to introduce local architectural regionalism, ethnic characters and multiple features from aspects of Yunnan-Tibet region's natural ecosystem and ethnic characteristics, meanwhile, combining with project analyses.

Erhai Resort Town backs against the Cangshan Mountain and faces the Erhai Lake, ascending from east to west with terraced fields scattering in the site. In order to build a primitive ecological resort, design gets inspiration from local terraced field and village. Enclosed terraced fields form village clusters, which seems that villages sit in terraced fields. The design restores landscapes to acquire primitive ecological view so as to echo the villages and the terraced fields commonly seen in Dali.

Grand Hyatt hotel is based on Naxi ethnic culture and integrates Lijiang beautiful snow mountain landscapes and Naxi ethnic flavor so that customers can experience local natural and ethnic characteristics.

InterContinental hotel is in accordance with modern aesthetic perspective and chooses butterfly as theme to present project's beauty in ecological, natural and elegant aspects, so it leaves customers an image of a colorful Yunnan.

Angsana Tengchong Hot Spring Village extensively shows Yunnan rustic scenery. Design modifies, restores and beautifies original villages from modern aesthetic perspectives to present a modern village-like holiday space featuring Yunnan cultural characteristics.

大理原生态度假风情 领略白族建筑文化

大理沧海一墅国际养生度假小镇

Dali Original Ecological Holiday Flair Experience Ba National Culture

Erhai Resort Town between Mountain and Sea, Dali

开发商：大理实力夏都置业有限公司　建筑设计：维思平建筑设计　建筑师：陈凌、隋鲁波、孙莹、曲克明、彭涵泽、周源、苗德、许浩等
项目地址：云南省大理白族自治州　基地面积：447 252平方米　建筑面积：393 572平方米　容积率：0.58　采编：张培华

Developer: Shili Group Dali Xiadu Properties Ltd. Architecture Design: WSP
Architects: Chen Ling, Sui Lubo, Sun Ying, Qu Keming, Peng Hanze, Zhou Yuan, Miao De, Xu Hao, etc. Location: Dali Bai Autonomous Prefecture, Yunnan
Site Area: 447,252 m² Building Area: 393, 572 m² Plot Ratio: 0.58 Contributing Coordinator: Zhang Peihua

苍山相靠，洱海相依，台阶状梯田散落，这为大理沧海一墅国际养生度假小镇的原生态度假风情提供了依据。设计从梯田建筑和民居村落中寻求灵感，将用地划分成若干块“梯田”建筑，通过围合而成的组团式的“村落”就坐落在这些“梯田”之上，以风景再生的方式，还原生态景观，与大理常见的景观——村落、梯田相呼应。同时以弧形屋面、横向纹理的片墙、格栅以及当地的石材等突显白族民族风情，与当地的民族性相契合，使建筑融入到大理这片土地中。

Erhai Resort Town backs against the Cangshan Mountain and faces the Erhai Lake, with terraced fields scattering in the site. In order to build a primitive ecological resort, design gets inspiration from local terraced fields and villages. Enclosed terraced fields form village clusters, which seems that villages sit in terraced fields. The design restores landscapes to acquire original ecological views to echo the villages and the terraced fields commonly seen in Dali. Arc roof, lateral texture slice wall, lattice window and local stones express Bai Nationality flavor, echoing vernacular ethnic characteristics. The project makes the resort perfectly blend in Dali.

大理沧海一墅-总平面 Master Plan

区位概况

项目地块位于大理古城与下关新城之间的七里桥感通寺区域，背靠苍山，面向洱海，地势西高东低，其间散落着台阶状的梯田，周围是一些小的民居聚落。东面距214国道300米，紧邻大理州交警培训中心。该地块属大理州、市严格保护的海西苍山坡地，西侧与国家地质公园相邻，自然环境及景观极其优越并且稀缺。

项目背景

大理传统的生活方式以民居聚落为主，传统的街巷一般不是直线，而是斜线、折线、曲线或者是顺应地势高低起伏，并且围合街巷的建筑都具有不同程度的凹凸变化，或者做了微小角度的偏转，这些都是传统街巷带给人们完全不同于城市居住区的空间感受。在建筑文化上，大理以白族文化为主，白族民居大多就地取材，以石头为主要建筑材料。

Location Overview

Erhai Resort Town locates at Qili Bridge-Gantong Temple area between ancient Dali and new Dali, against Cangshan Mountain and in front of Erhai Lake, and it is adjacent to the National Geological Park in the west. It is 300 m away from the 214 National Road in the east and very close to the Dali Prefecture Traffic Police Training Center. This site, enjoying favorable and scarce natural environment and landscapes, belongs to Haixi Cangshan Mountain sloping fields which are strictly protected by Dali Prefecture.

Project Context

Dali traditional lifestyle is in vernacular dwelling settlements. Traditional streets are in slash, fold line or curve line distribution or even along with undulating terrain instead of straight line. The buildings, enclosed by streets and lanes, have a convex-concave variation or some subtle deflection angles, which brings us a totally different feeling from urban living space. Bai nationality culture dominates in Dali, and Bai ethnic people are adept at utilizing local materials, so stone becomes a dominant material in their architecture.

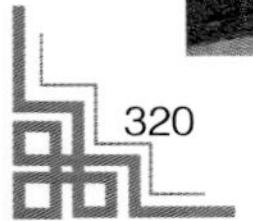

设计理念：村落组团

项目定义的村落式组团保留传统生活空间的街巷感受，组团内部的建筑有规律地错动，适当地在组团内部留有一处空地，作为组团内的公共交往区域。组团和组团之间在色彩和材质上既有联系又有所区分，使每个组团各具特色。

规划布局：梯田概念

项目设计着重考虑“如何用利用现状的地形，如何把大理的文化融入进来”。设计中融入“梯田”“村落”这两个概念，通过对场地的重新整合，疏导高差关系，审慎定义每个组团村落的规模，使之与地形契合。并将用地划分成若干块“梯田”建筑，通过围合而成的组团式的“村落”就坐落在这些“梯田”之上。

Design Conception: Village Cluster

The village cluster strives to retain a traditional lane space feeling, so the buildings in the cluster are arranged in an orderly dislocation, while reserves a vacant lot as a public communication area. Among the clusters, the application of color and material is related and distinguished somewhat, remaining every cluster with its own characteristics.

Planning Layout: Terraced Field

The project lays great emphasis on how to use terrain features and integrate Dali culture in design, so “terraced field” and “village cluster” concepts are brought into the design. The site is re-integrated, and discrepancy in elevation is adjusted and even every scale of village cluster is defined to fit the terrain. The enclosed terraced fields form village clusters, like villages sitting on terraced fields looking from far away.

小贴士

白族崇尚白色，因此白族民间建筑以白色为主色调，墙面以石灰粉刷。白族民间建筑一般就地取材，广泛运用石头，墙角、门头、窗头、飞檐等部位采用具有几何线条和麻点花纹的石块（条），墙壁常用天然鹅卵石堆砌。

Tips

Bai nationality admires white color that is the dominant hue in architecture with lime walls. Bai nationality architecture usually draws on local stone material. Geometrical lines and pocking or strip pattern stones are used on corner, door head, window head and eave, and walls are piled up by natural pebbles.

建筑设计：演绎白族文化

大理以白族文化为主，如何提炼传统文化的元素与现代建筑相融合是设计需思考的课题。屋顶是一个地域具有代表性的符号，建筑的坡屋顶是由白族民居的布瓦联想而来，造型新颖飘逸又不失传统韵味。

立面设计上，简洁轻巧的弧形屋面与大理传统民居相呼应，横向纹理的片墙组合打破立面的呆板格局让建筑整体感更为轻盈。庭院的入口雨棚与坡屋顶相呼应，入口门的设计上引入传统格栅式样，通透的格栅透露出庭院内的绿意盎然，让石材砌筑的院墙活泼了起来。

在建筑的选材上，就地取材，充分考量当地的地域文化特色，房屋多以石砌为主，屋顶选择石板瓦。立面上浅灰色横向纹理的毛糙质感涂料给人以视觉上的冲击，极具民族风情。外围山墙上的不规则拼接的石材效果，自然原始，暖色石材与灰色石材的组合极富亲和力，与当地的传统民族的石头房子相呼应，让建筑更接“地气”，更好地融入到大理这片土地上。

景观设计：风景再生

项目在设计之初，就将建筑与场地的关系紧密地结合在一起，最大限度地减少土方开挖量的同时也有效地利用原有地势的优势，让户户观海成为可能。山为阳刚，水为阴柔；建筑为阳刚，绿景为阴柔。在山水阴阳和谐共生的自然环境中，新的建筑与绿化环境也是互相交融熠熠生辉，通过与自然的结合，实现风景的再生。

Architecture Design: Bai Nationality Culture Performance

Bai Nationality culture has dominated Dali for a long history. How to refine traditional culture elements and integrate them in contemporary architecture is a key issue in design. Usually roof is a representative symbol in a district. The pitched roof design is inspired from the roof commonly seen in Bai Nationality villages. The modeling is fresh, with a traditional flavor.

Lithesome and concise arc roof echoes traditional Dali dwellings, and lateral texture slice wall breaks out rigid layout on facade, making the overall architecture looks light and graceful. The canopy in the yard entrance echoes pitched roof, and the door is in traditional lattice form. Vibrant green in the yard stretches out from the lattice window, enlivening the stone yard walls.

Material selection fully considers local cultural features, so house adopts stone as main material, and roof covers with slating. Rough lateral texture light-gray coating painted on surface wall stimulates visual impact, bringing full of local ethnic flavor. The peripheral gable walls are wrapped with irregular patchwork stone gobbets, natural and primitive. The combination of warm color stone and gray stone emits obvious affinity, corresponding to local traditional stone houses. The architecture design boots on the ground and is naturally blended in Dali.

Landscape Design: Scenery Revivification

In the beginning of the design, the relation of architecture and the plot is closely united so as to reduce earthwork excavating volume and take advantage of the topography, so seascape is available for every family. Architecture like mountain stands for masculinity and green scene like water represents femininity. In a harmonious and natural environment, new buildings and green environments blend with each other. Through reasonable combination with nature, the design realizes scenery revivification.

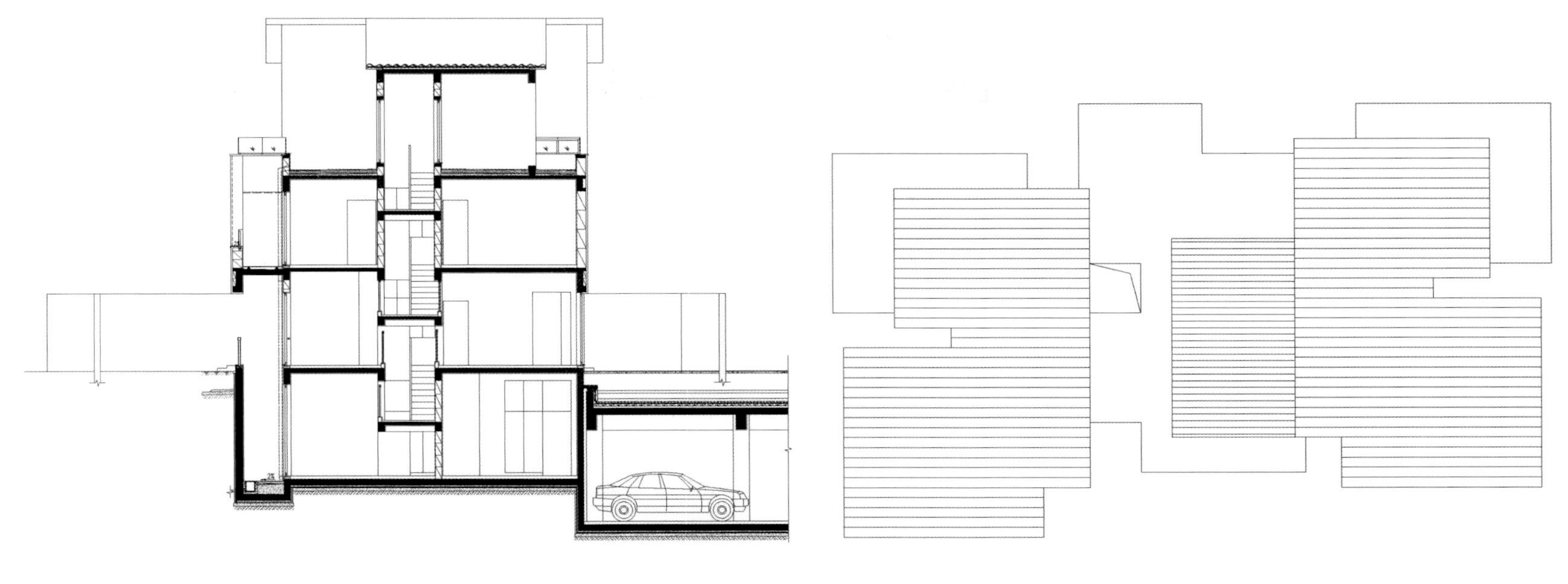

剖面图 Section

屋顶平面图 Roof Plan

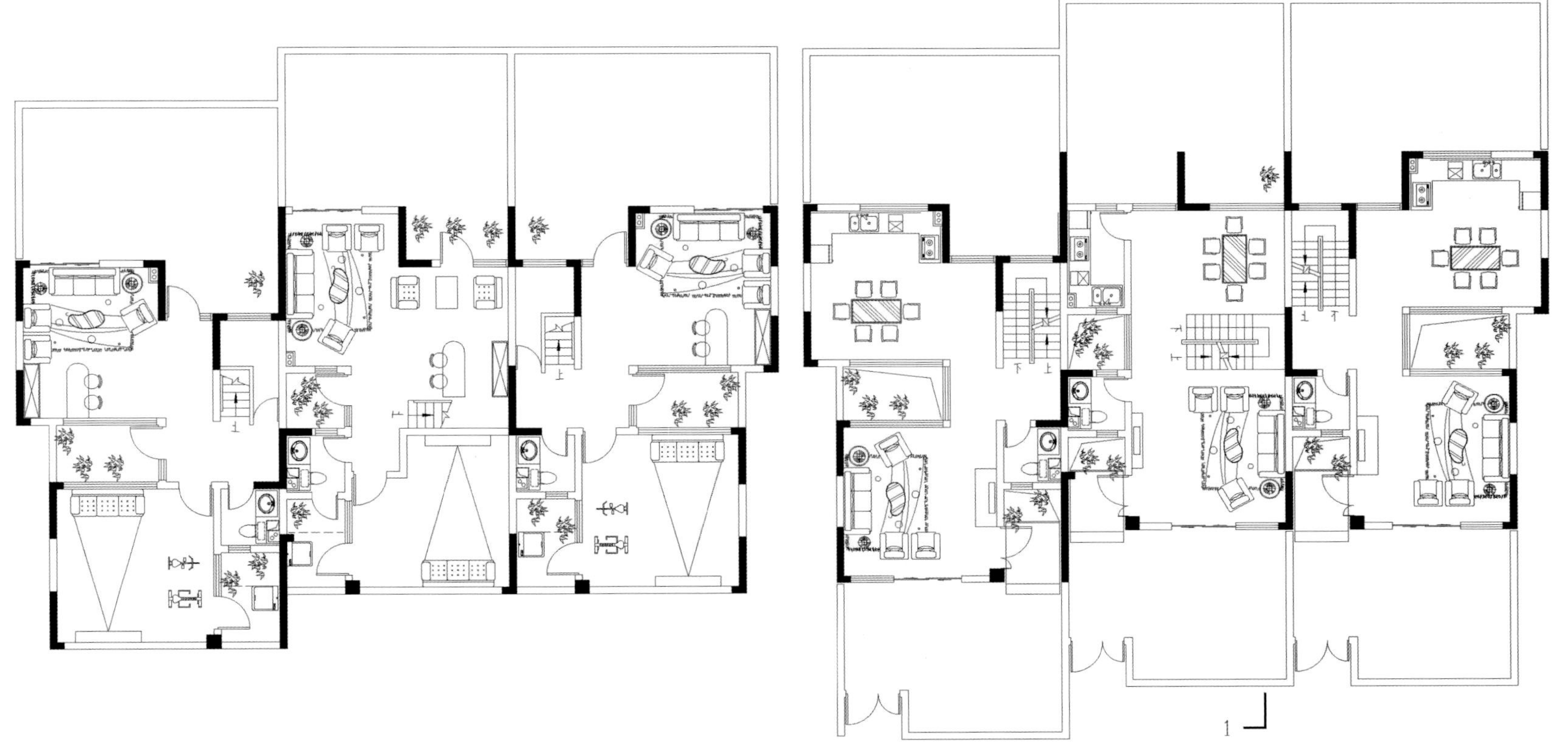

负一层平面 -1F Plan

一层平面 1F Plan

户型设计

单体平面布局中，功能动静分区明确，层层递进的设计空间满足生活和心理需求，所有单体均有两个庭院：入户庭院和相对私密的生活庭院。房屋在入口设有玄关，入户层为起居与厨房餐厅公共区域，地下室是多功能活动区，房间环抱的采光天井直通地下，让所有房间均有直接的对外采光和通风，居住空间在二层和三层，屋顶书房和前后两个露台相连，在休息时，可观山望水。

House Type Design

Monomer plane functional layout has clear living and private zone separating. Layer-upon-layer progressive space satisfies living and psychological needs. Every monomer equips with two courtyards: the entrance courtyard and relatively private living courtyard. Stepping through vestibule entering the first floor is living room and public area such as kitchen and dining room. The basement is set as a multi-functional room. Lighting patio, surrounded by bedrooms, sinks to basement, which brings every room with lighting and fresh air. The second and the third floors are for bedrooms, the study on the roof connects to front and rear terraces for inhabitants to enjoy landscapes.

云南乡村特色 现代美学视角

云南腾冲悦椿温泉村

Yunnan Village Characteristic Modern Aesthetic Perspective

Angsana Tengchong Hot Spring Village, Yunnan

业主：云南省城市建设投资集团有限公司 开发商：腾冲玛御谷温泉投资有限公司
建筑/室内设计：北京阿其德尼设计咨询有限公司 景观设计：美国SWA集团 项目地址：云南腾冲 主创设计师：王广成
项目面积：60 000多平方米 建筑面积：19 104.38平方米 采编：谢雪婷、陈惠慧

Proprietor: YMCI Developer: Mary Imperial Valley Hot Springs Tengchong Investment Co., Ltd.
Architecture Design/ Interior Design: Beijing Arkiteknique Design Consulting Co., Ltd.
Landscape Design: SWA Group Location: Tengchong, Yunnan Chief Designer: Wang Guangcheng
Site Area: over 60,000 m² Building Area: 19,104.38 m² Contributing Coordinator: Xie Xueting, Chen Huihui

腾冲优越的自然资源及深厚的文化遗产，自然村落气息浓厚，为项目设计提供了设计灵感。项目设计从基地现状出发，对原有村落以现代美学角度进行修饰、恢复和美化，保有原有的乡村魅力，以飞檐建筑、鹅卵石小径等元素，演绎出富有特色的云南文化现代村落度假空间。

Tengchong County enjoys advanced natural resources and opulent cultural relics. Inspired by local natural village breath, the project design combines with local site conditions to modify, restore and beautify existing villages. Tall eave and cobblestone lane elements reflect a modern resort space of Yunnan Characteristics.

总平面图 Master Plan

区位分析

腾冲悦椿温泉村坐落在玛御谷之中，距腾冲镇约15公里，距离腾冲机场约30分钟车程。项目周边15分钟车程的范围内，有几处非常知名的景点，比如腾冲湿地与叠水河瀑布。除了温泉，地热活动还在该地区形成了中国最大的火山群。宗教名山云峰山距离温泉村约1小时车程。现有约15个规模不等的村落位于项目地块内及紧邻项目地块周边的边界区域内。

项目背景

腾冲位于云南省西部，自西汉起几经更迭，是著名的侨乡、文献之邦和翡翠集散地，亦是省级历史文化名城，居住有汉、回、傣族等20多个民族，拥有丰富的文化遗产。腾冲还是世界罕见并且是最典型的火山地热并存区域，有99座火山，88处温泉。

业态功能

项目拥有腾冲最大的室外温泉浴场，内设有43个饮用水级别的矿物质温泉汤池、配备会议室、麻将室、电子游戏室、台球室、乒乓球室、儿童俱乐部、餐饮等配套设施。其中环境一流的会议室面积达100平方米，设施齐备，满足40人的中小型会议需求。此外，还配有湿蒸房、足疗室、SPA理疗室与贵宾更衣室，满足高品位客人的各式需求。

规划设计

鉴于现有约15个规模不等的村落位于项目地块内及紧邻项目地块周边的边界区域内，且这些村落现有的建筑现状和风格大相径庭，从具有地方建筑风格的传统庭院住宅到当地政府机关所在的现代混凝土结构建筑物都有。设计通过对村落的建筑物进行修饰、恢复和美化，以便能够成功地将项目地界范围内的村落与整体开发项目有效融合，或至少将那些在未来可能会对拟建开发项目的氛围造成的任何负面影响减小到最低程度。

Location Overview

Angsana Tengchong Hot Spring Village locates in the Mary Imperial Valley, 15 km to Tengchong County and 30 min drive to the Tengchong Airport. In a range of 15 min drive, there are several famous scenic spots, for instance the Tengchong Wetland and the Dieshuihe Waterfal. Except hot spring, local geothermal activity produces the largest volcanic cluster in China. A religious mountain the Yunfeng Mountain is 1 hour away from the hot spring village. Until now, there are almost 15 various scales villages in the site or near the site.

Project Context

TTengchong County situates at west of Yunnan with a history dating back to the Western Han Dynasty. It is a famous home town of overseas Chinese, a literature state and a jadeite distributing center. Now it is a historical city with more than 20 nationalities, such as Han, Hui and Dai nationalities, enjoying rich cultural relics. Tengchong is a rare and typical area coexisting volcano and geothermal activity, with 99 volcanoes and 88 hot springs.

Services

The project has the largest outdoor hot spring resort in Tengchong, including 43 mineral spring bathing pools which are up to the standard of drinking water level. In the resort village, conference room, mahjoon room, electronic game room, billiard room, Ping Pong room, kid's club and F&B supporting facilities are all ready. The conference room equips complete facilities and has first-class environment of 100 m² for small or medium-sized meetings, allowing for 40 people. In addition, wet steam room, foot massage room and Spa room, guest changing room provide at any time to meet high grade tourists' demands.

Planning Design

In view of 15 various scale villages in the site and adjacent plots, and local traditional courtyard residence and modern concrete structure government agencies coexisting yet so different from each other in architectural styles, and these villages appearing in various building conditions, the design modifies, restores and beautifies some villages to integrate them with the general project. At least, the planning design aims to minimize potential negative effects on the atmosphere for the future development projects.

酒店建筑外观大堂 Hotel Appearance

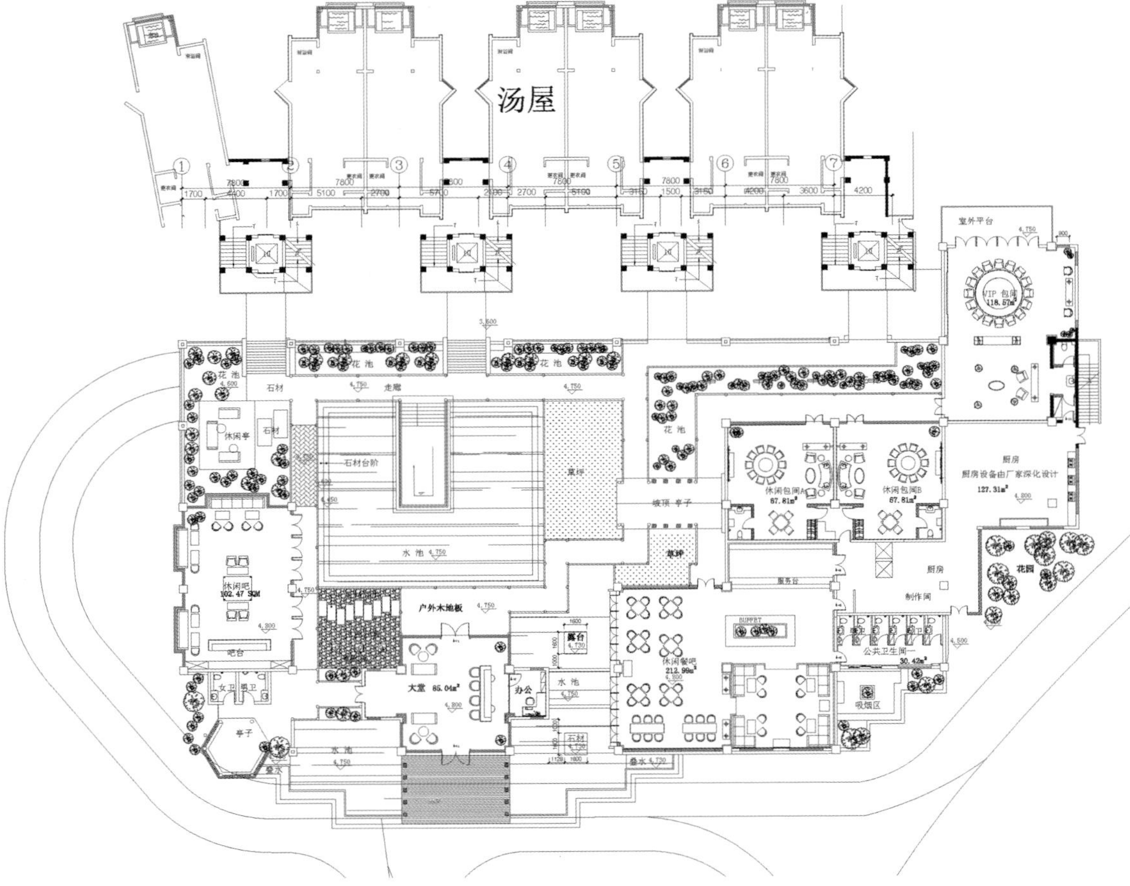

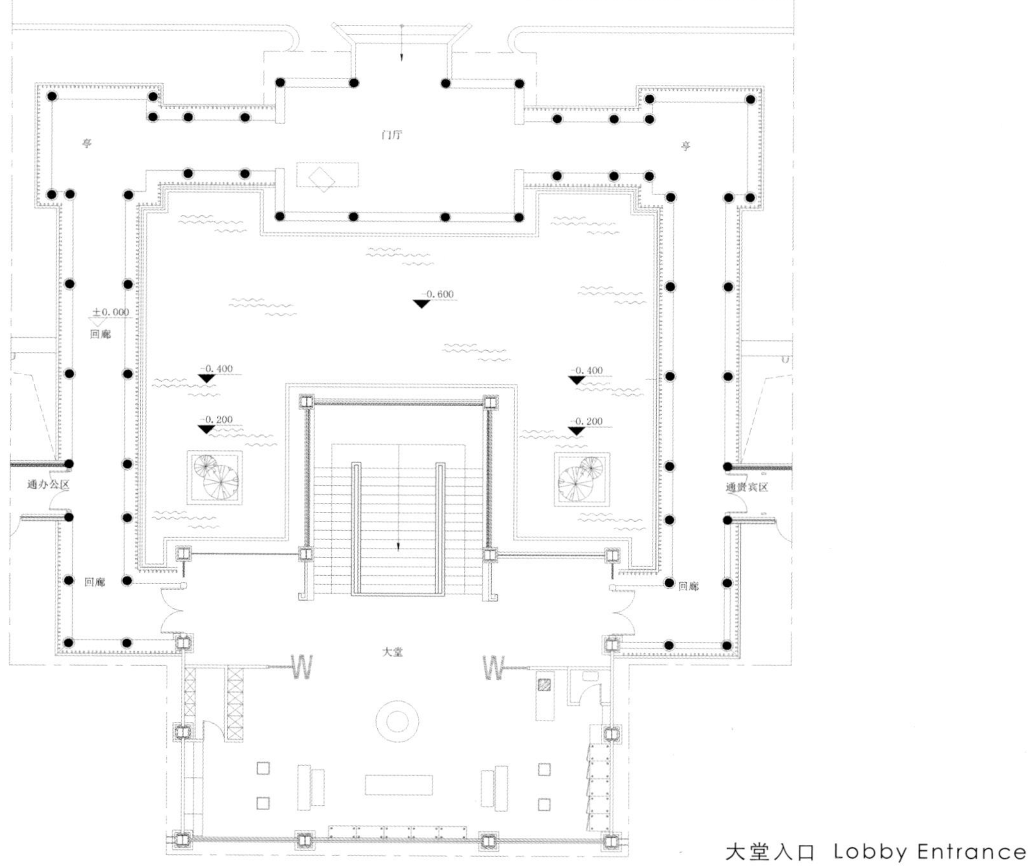

大堂入口 Lobby Entrance

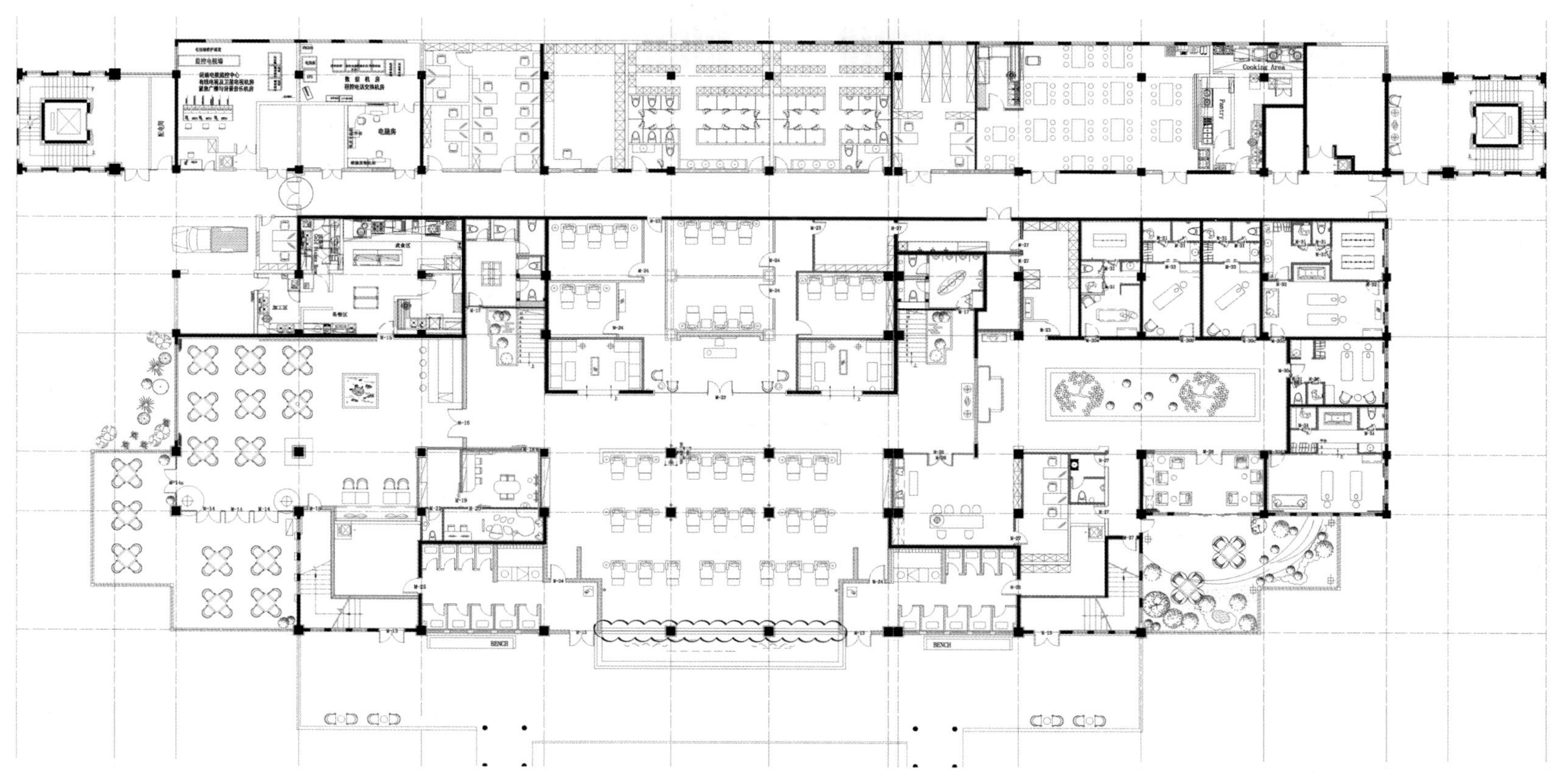

温泉中心底层 Hot Spring Center Ground Floor

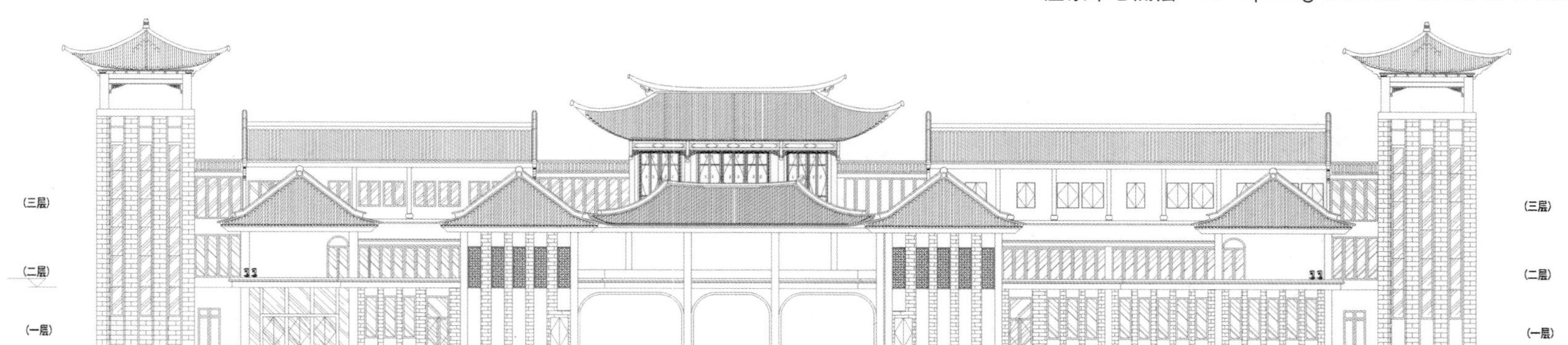

温泉中心会所外立面 Hot Spring Center Club Elevation

豪华汤院外立面图 Hot Spring House Elevation

建筑设计

项目设计风格精致独特，建筑设计以传统中国诗歌绘画为灵感，将现代与乡村魅力风格相融合，充分展示该地区丰富的文化传统。高挑的飞檐建筑和石板小径在精致的花园中蜿转延伸，如诗如画。项目拥有37栋精美汤屋及汤院，呈现出时尚现代的装修风格。温泉中心采用充满东方魅力的建筑外观，以不同的接待区域为特色。

Architectural Design

The project has delicate and unique style that the architectural design takes traditional Chinese poem and painting as design inspiration, and combines modern style with village characteristics, which expresses local rich cultural traditions. Tall eaves and stone lanes bend and extend in beautiful gardens, like a Chinese poem or an ink painting. The project contains 37 delicate bathing pool houses and courtyards, all in modern decoration style. The hot spring center applies eastern style facade, flaunting with various reception areas.

蝴蝶主题设计 七彩云南气息

昆明洲际酒店

Butterfly Theme Design   Colorful Yunnan Ambiance

InterContinental, Kunming

开发商：云南省城市投资建设有限公司　管理集团：洲际酒店集团　室内设计：CCD香港郑中设计事务所
项目地址：云南省昆明市　项目面积：约100 000平方米　采编：谭杰

Developer: YMCI Group　Property Management: InterContinental Hotels Group
Interior Design: Cheng Chung Design　Location: Kunming, Yunnan　Site Area: about 100,000 m²　Contributing Coordinator: Tan Jie

七彩炫丽是云南留给世人的印象之一，原生态、古朴的气息是其所具有的属性之一。而蝴蝶所象征的美丽、优雅、生态和自然的特点以及其七彩的翅膀，在无形中，便与云南产生了一种共鸣。昆明洲际酒店以“隽美蝴蝶”为设计主题，于细微之处演绎出绚烂、自然的建筑空间。同时，将云南传统特色及国际时尚元素相融合，赋予项目独特的地域体验。

Colorful and magnificent image is one of impressions of Yunnan, and original and primitive breath is also one of its characteristics. Butterfly symbolizes beauty, elegance and nature, so its colorful wings just echo the beauty of Yunnan. InterContinental takes butterfly as design theme to create a splendid and natural space, and meanwhile, Yunnan architectural characteristics and international fashionable elements are well combined to form a unique regional experience.

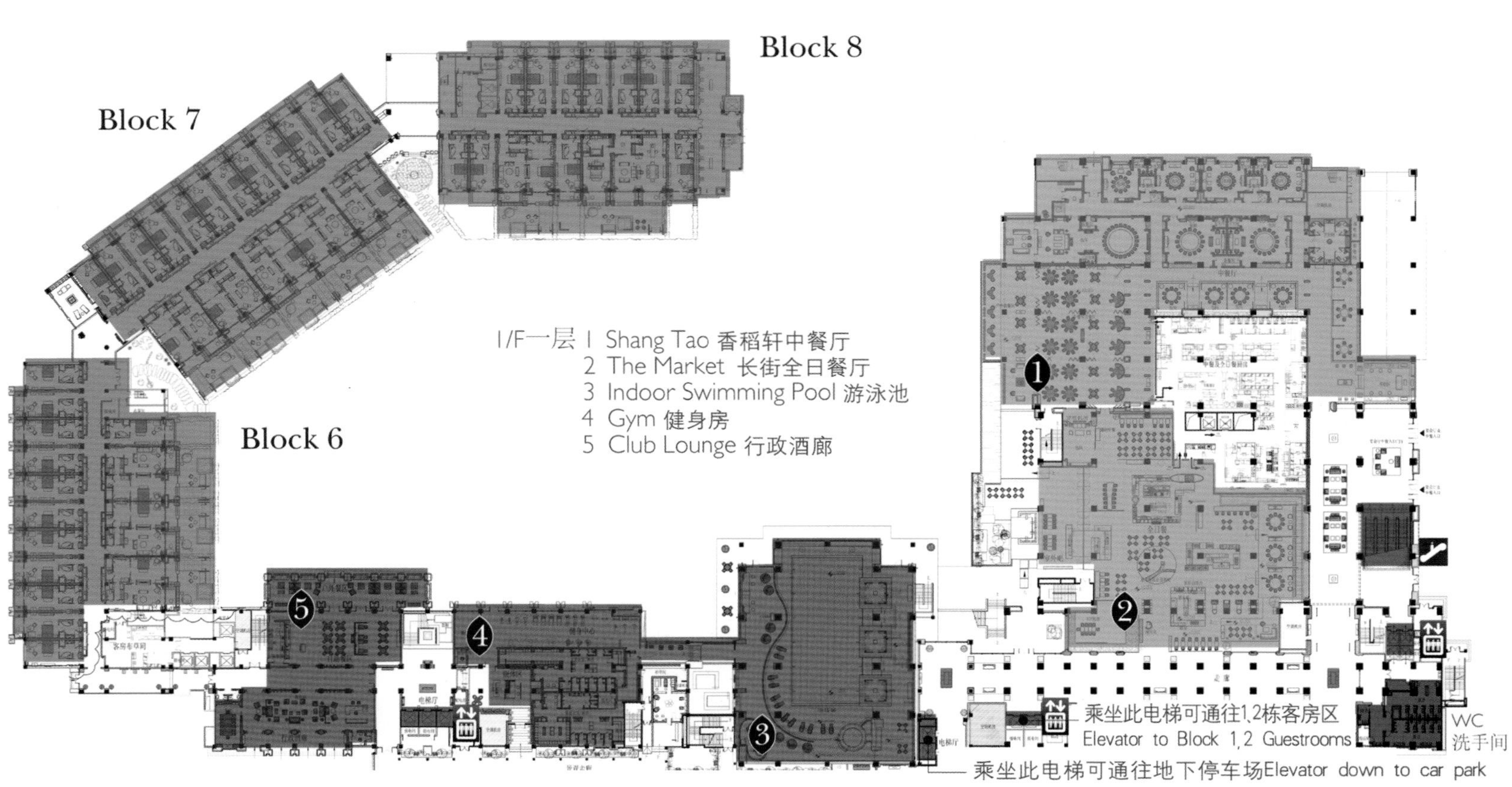

酒店平面图-1层 -1F Plan

区位图 Location Map

区位分析

昆明洲际酒店位于云南省昆明市滇池国家旅游度假区，毗邻“高原明珠”滇池湖畔，距市区仅20分钟车程，驱车40分钟即达昆明长水国际机场。酒店周围山、水、林、园皆备，与西山森林公园、云南民族村等著名旅游景点互相依托，再加上与市中心CBD地区仅咫尺之遥，使得酒店处于一个商务、游览、娱乐、度假俱佳的极好地理位置。

项目背景

云南省，简称是“滇”或“云”，是人类文明的重要发祥地之一，有彩云之南、七彩云南之称。省会昆明三面环山，由于地处低纬高原而形成“四季如春”的气候，特别是有高原湖泊滇池在调节着温湿度，使这里空气清新、天高云淡、阳光明媚、鲜花常开。

项目概况

酒店拥有7栋6层楼高的客房，其中6栋简洁古朴的建筑错落有致地镶嵌在酒店的亭台园林之间。541间设计独特新颖的高级、豪华客房及典雅套房，面积从110平方米到265平方米不等，每间客房均有可由客人自主开关的窗户。酒店拥有超大多样的会议场所，1 600平方米的云南大宴会厅是目前昆明高端酒店中最大的宴会厅，可同时容纳1 200位客人；8间会议室布局灵活，可满足各种规模的会议需求。水疗中心和健身设施应有尽有。此外，酒店还设有5个餐饮设施，分别为长街全日西餐厅、香稻轩中餐厅、香堡法餐厅、彩蝶吧、行政酒廊。

Location Overview

Sited in Kunming Dianchi National Tourist Resort, the hotel is close to Dianchi Lake which is renowned as a pearl on plateau, and just 20 min drive to Kunming downtown, and 40 min drive to Kunming Changshui International Airport. Surrounded by mountains, rivers, forests and gardens, the hotel, Xishan Forest Park and Yunnan Ethnic Village serve as a foil to one another. In addition, the hotel is throw stones away from Kunming CBD, which enjoys excellent site for business, traveling, entertainment and vacation.

Project Context

Yunnan Province, shortened for Dian or Yun, is one of an important birthplaces of civilization. It is also called the South of Colorful Cloud or Colorful Yunnan. Kunming, the provincial capital, is surrounded by mountains at three sides. Because it lies on a plateau in the low latitudes, here enjoys spring climate in four seasons. In addition, there is a plateau lake-Dianchi Lake to adjust temperature and humidity, which makes Kuming have clear air, serene sky, bright sunshine and flowers blooming at all seasons.

Project Overview

The hotel includes 7 six-storey-high guest room buildings, and 6 of them scatter in gardens of the hotel, succinct and vintage..There are 541 high grade, deluxe and elegant suites with unique design arranging from 110 m^2 to 265 m^2, and every guest room equips windows can be opened freely by guests. It has super-sized multifunctional conference rooms, and the Grand Yunnan Banquet Hall of 1,600 m^2 is the largest one in Kunming high grade hotels, which is available for 1,200 guests. 8 conference rooms are flexible in layout so as to satisfy requirements of every scale meeting. Hydrotherapy center and fitness facilities are always ready for guests. Moreover, there are 5 F&B spaces: a Long Street Western Dining Hall for 24 hours, a Shang Tao Chinese Restaurant, a Xiangbao French Dining Hall, a Colorful Butterfly Bar and a Club Lounge.

建筑设计：蝴蝶主题

酒店设计关注对云南文化的展现，选材也体现出贴近自然的倾向。整体设计以蝴蝶为主。蝴蝶，代表了美丽、优雅、生态、自然，这是品牌对当地风情深入洞悉的结果，蝴蝶七彩斑斓的翅膀也如同云南的七彩浮云，炫目多彩，很好地诠释了酒店的独特文化特征。

室内设计

酒店客房设计融汇了云南传统特色及国际时尚元素，光影下栩栩如生的蝴蝶壁画和充满大自然气息的设计新颖独特。大堂设计符合蝴蝶主题，透露着文化气息，展示出云南的独特魅力。行政楼层设计尊贵典雅，既有精心设计的室内酒廊，也有行政专享的室外露台，客人可将酒店园林美景尽收眼底。所有会议室的命名均取词于素有中国“古今第一长联”之称的昆明“大观楼长联”中对美景的描绘，内涵深刻且与会议室的装饰相得益彰。

Architectural Design: Butterfly Theme

The design emphasizes on revealing Yunnan Characteristic cultures, so the decoration materials prefer to nature. The major design theme is butterfly, which means beauty, elegance, ecology and nature, and certainly, it is the result of deep insight in local customs. The multicolored wings of butterfly just like colorful clouds, glaring and gorgeous, well interpreting a special culture of the hotel.

Interior Design

The guest room design integrates Yunnan traditional characteristics and international modern elements. Vivid butterfly murals under shadow and full of natural breath design are original. The lobby design also echoes butterfly theme, emitting out cultural breath and Yunnan peculiar appeal. The administrative floor is noble and elegant with exquisite interior lounge and special exterior terrace for guests to enjoy the beauty of the garden in the hotel. The name of every conference room is from the depiction in Kunming “Daguanlou long Chinese antithetical couplets” which are renowned as “the longest Chinese antithetical couplets from ancient” in China. The deep connotation of the names complement well with the interior decoration.

古城慢生活 纳西民族风

丽江金茂君悦酒店

Slow Life in Ancient City　Naxi Ethnic Style

Grand Hyatt, Lijiang

开发商：方兴地产（中国）有限公司　室内设计：P49
室内设计师：CHAKKRAPHONG MANIPANTI　项目地址：云南丽江　项目面积：约82 063平方米　采编：陈惠慧

Developer: Franshion Properties (China) Limited　Interior Design: P49 Deesign
Interior Designer: Chakkraphone Manipanti　Location: Lijiang, Yunan
Site Area: about 82,063 m²　Contributing Coordinator: Chen Huihui

优美的丽江雪山风光、淳朴的纳西民族风情，丽江金茂君悦酒店置于其间，不仅营造出融于自然的空间，更是将纳西族的特色文化与建筑特色与现代时尚元素相融合，通过原生态吊灯以及精致的手工制作设计元素等细节设计，让旅客感受到丽江的民族风韵以及深厚的文化底蕴，营造出属于丽江的、富有慢生活气息的一片纯净空间。

Situating in beautiful landscapes of Lijiang snow mountain and surrounded by rustic Naxi national customs, the Grand Hyatt hotel creates a space integrating in nature, and meanwhile it combines Naxi ethnic cultural and architectural characteristics and modern fashionable elements. Original droplights and delicate handmade ornaments present design details that makes people feel local ethnic flavors and profound cultural foundations. The hotel creates a pure space especially belonging to Lijiang and featuring in slow life characteristics.

区位分析

项目毗邻丽江旅游核心地带——玉龙雪山区域及古城区原世界遗产公园，紧邻束河古镇，毗邻玉龙雪山高尔夫球场。酒店交通便利，距离丽江三义机场45分钟车程，距火车站40分钟车程。

项目背景

丽江，是一座拥有800年历史的联合国教科文组织遗产城市，纳西文化在这座城市被完好地保存着，传统的纳西建筑和生活方式延续至今。纳西族古文化——东巴文化，是中国灿烂的民族文化中的一部分。

设计主题

项目的主要设计主题是将丽江当地沉淀了几个世纪的文化元素融合在一起。酒店的每个区域都有不同的局部设计特征，使得它不仅起到分区的功能，还模拟了整个“城市”的社会互动。

规划设计

项目坐落于金茂雪山语综合项目内，整座酒店环湖而建，巧妙地隐于风景之中。酒店拥有312间豪华客房、套房及别墅，4间融合当地特色的餐厅、茶室和酒廊，3 000平方米的多功能会议及宴会设施包括900平方米的金茂宴会厅和10个富有独特风格及自然光的君府多功能会议室以及“净”SPA水疗设施。

Location Overview

Grand Hyatt hotel is close to the core tourist belt in Lijiang where locates the Jade Dragon Snow Mountain, the World Heritage Park of Lijiang in ancient city area, the Shuhe Ancient Town and the Golf Course of Jade Dragon Snow Mountain. The beautiful hotel enjoys convenient transportation conditions, just 45 min drive to the Lijiang Sanyi Airport and 40 min drive to the local Railway Station.

Project Context

Lijiang has more than 800 years' history, and it was designated by UNESCO to be a heritage city, because Naxi ethnic culture is perfectly preserved here, and Naxi architecture and life style have been continued up to now. Dongba culture is the fruit of Naxi culture that has been inherited from generation to generation, which also is a splendid ethnic culture part in China.

Design Theme

The Lijing Grand Hyatt situates in the core area of the Lijiang Ancient Town, hence design theme aims to integrate local cultural elements precipitated for centuries. While every section in the hotel possesses unique detail features, which acts as sectional partition and imitates the social interaction of a complete “city”.

Planning Design

Locating in the villa project of Jinmao Snow Language, the Grand Hyatt hotel builds by lake, hiding in beautiful landscapes. The hotel has 312 deluxe guest rooms, suites and villas, 4 local featured dining halls, tea house and lounge, 10 featured multifunctional conference rooms with a total area of 3,000 m², banquet halls of 900 m² and SPA facilities.

酒店布局图 Functional Distributing Map

小贴士

纳西民居建筑一般为穿斗式结构、垒土坯墙、瓦屋顶，设有外廊（厦子），三坊一照壁是丽江纳西民居最基本最常见的民居形式。建筑上端深长的“出檐”是具有一定的曲度的面坡，避免了沉重，突出柔和优美的曲线。外廊是纳西民居中的一大显著特点，供居住者会客、休闲使用。

Tips

Naxi ethnic dwellings construct in bucket formation with adobe wall, tiled roof and outer veranda. Three squares with a screen is the most basic and common dwelling form in Naxi ethnic architecture. The top eaves are surfaces with curvature that highlight gentle and soft curves and avoid heaviness. The outer veranda is a spotlight in Naxi dwellings for the use of residents to receive visitors and enjoy leisure.

建筑设计

项目客房设计巧妙融合了现代时尚元素和纳西民族建筑特色的精华，呈现豪华的精致质感生活。

Architectural Design

The guest room design integrates modern fashionable elements and Naxi ethnic architectural quintessence to present a sumptuous and exquisite life.

室内设计

设计以丽江古镇鲜明的本地文化色彩为契入点，将酒店的室内设计分区列表，然后尝试将当地不同的文化元素小心地应用到每个区域，比如当地的街头小吃、纳西语言文字、吉祥物、古镇的商铺氛围等，最大可能地让客人获得最真实的当地体验。

客房

客房采用了木门板，不仅使房间感觉更加温暖，而且模仿了本地房间的建筑特点。手工编织的藤条墙布和墙上的挂毯使得整体氛围更加柔和，并且起到装饰作用。房间内有从现代到古典的不同种类和风格的家具。枕头采用当地的手工织物，为房间的色调增添亮点。

大堂：原生态吊灯

大堂充满了贵族气息，铺石和仿古木地板分隔开了房子内部的空间，以符合当地建筑的特点，并且避免了墙壁的遮挡。一系列的多个纳西风格的门板被用来划分空间。大厅中央座位的上方，挂着一个原生态的吊灯，其形状仿照的是当地民族使用的牛铃。牛铃柔和的声响，是当地民族十分熟悉的声音，这样与当地生活息息相关的元素也特意被引入到酒店设计中。

简约现代风格的家具放置在传统的东巴建筑空间中，在反差对比下形成独特的美感。仿古铜作品系列的小配饰，充分展现了酒店的度假风格。柔软舒适的藏式风格的地毯随机放置在大堂中央座位周围，与一些地方手工编织的纺织品和靠枕搭配在一起。

Interior Design

The interior design highlights in absorbing Lijiang vernacular cultural elements to build divisions. Various divisions possess different cultural elements, such as street snack, Naxi ethnic language, mascot and ancient town shop. The arrangement brings people a real experience of local life.

Guest Room

Guest room design imitates local room features, adopting wooden door to make the room feel warm and handmade rattan wall covering and tapestry to soften the interior ambiance and decorate it in the mean time. Furniture combines modern and classical styles. Local bright-colored pillows lighten the whole color of the guest room.

Lobby: Original Ecology Droplight

The whole lobby boasts a noble breath. The interior space is divided by band stone and antique wood flooring that meet local architectural characteristics and prevent wall sheltering. A series of Naxi ethnic style door planks are as partition in the space. On the central of the ceiling decorates an original droplight whose shape is like a local cowbell. Because the cowbell voice is a very familiar sound in local, which brings local life elements to the hotel naturally.

Under the contrast of minimal modern furniture in traditional Dongba architectural space creates a unique sense of beauty. A set of antique brass ornaments express holiday atmosphere, and Tibetan style carpets are paved around central seats, matching with handmade textiles and ethnic cushions.

全日餐厅：街头文化

酒店全日餐厅的创建仿佛是街头小吃的嘉年华，餐厅没有围墙，各色美食一览无遗。

丽江木质镂空货架和石材、仿古铜、瓷砖等不同材质的墙面，将每种地方美食划分了区域，但现场烹饪台和食品展示柜被放置在同一条石头铺就的通道上，使人有走在丽江老城区的感觉。部分裸露的天花板可以看到屋顶的结构，一些手工制作的木质、竹编或仿古铜的小吊坠，为空间增添了温暖的感觉。在全日餐厅，大街上常见的食材充当了展现当地风格精髓的关键因素。装修材料注重触感，主要集中在旧木材、仿古铜和石裂面（五色石）等。简单的现代家具和古典风格的家具混合搭配，一些古董柜和栗色的椅子、门面板和炊具、食品调料的容器，共同营造了丽江街头美食的气氛。

中式餐厅：镂空式屏风

中式餐厅被打造成丽江老城区主干道两旁的商铺概念。从“街道”这一侧看过去，穿过美丽的湖景，便是餐厅的用餐区域，视野十分通透。丽江风格的镂空式屏风和各种装饰物，为客人提供了亲密、温馨的就餐体验。

连接私人用餐区和主用餐区的，是各种装饰艺术品。包间风格典雅，配有左右对称的多种丽江木花纹的镂空设计，家具的设计灵感则来源于东巴家具和现代家具的融合，给人以古朴优雅之感。

All Day Dining: Street Culture

The all day dining in the hotel is like a carnival of street snacks, and the dining halls have no walls, so all sorts of food are fully flared.

Multiple wall materials, such as wooden hollow shelf, stone, antique brass and ceramic divide all kinds of local cuisines, and cooking table and food cabinet are on a same stone path, which makes people like stroll in the Lijiang old town. Roof structure can be seen from part uncovered ceilings, and some handmade wooden, bamboo and antique brass pendent ornaments bring a warm feeling in the hotel. In the all day dining halls, common street snacks are the core to present local essence. As for decoration materials, it focuses on old wooden materials, antique brass and rock fracture (variegation stone). The mix-match of modern and classical furniture, antique cabinets, nut-brown chairs, door planks, cookers and containers of food seasoning together create a Lijiang street snack atmosphere.

Chinese Restaurant: Hollow Screen

The Chinese Restaurant is built like a shop on the trunk road in Lijiang Old Town. Seen from the “street”, dining area is just across a lake with pass-through visions. The local style hollow screen and other ornaments offer an intimate and comfortable dining experience.

Opulent artistic ornaments connect private dining areas and the main dining area. Private rooms are elegant with symmetric hollow wooden carvings. The furniture design is inspired by the combination of Dongba and modern furniture, bringing a pristine and elegant feeling.

大俱乐部

简奢的门板设计细节体现出非同一般的沉稳大气。中式陈列柜中的古董艺术品则会增加一些家的氛围。该空间的主色调是大地色系，同时采用靛青蓝和银色两种鲜明的色彩。

SPA

从主大厅出来，走过一条鹅卵石路，便来到了水疗中心。水疗中心与中餐厅一样，同样被打造成老城区街道两侧的一个商铺。“香草和香料”零售店也被引入到了水疗中心。抵达水疗馆接待处的架子和抽屉中展示着装着各种药草和香料的容器。货架以屏风的形式将各组座位隔开，保证客人的私密性。接待岛台采用朴实的色调、纹理饰面墙和木质镂空设计。

茶室

茶室墙上挂着的丽江当地宣纸画，讲述了茶的悠悠历史。家具结合了古老的中式、东巴——纳西和现代家具的风格，以满足客人在雪山湖水间品茶的雅兴。

Grand Club

Concise yet luxury door plank design details express extraordinarily calm and imposing manner. Antique artworks display in Chinese style cabinets, bringing an ambiance of home. Earth tone is as the dominant hue, while some indigo blue and silver colors are added in parts to lighten the Grand Club.

SPA Treatment Room

Walking out of the lobby and through a pebble path arrives a SPA center. Like the Chinese Restaurant, the SPA Treatment Room is built to be a shop standing on the trunk road in Lijiang Old Town. “Herbs and spices” retail shop is arranged in the SPA center. In front of reception table is a row of shelters with drawers, displaying all sorts of herb and spice containers, and the shelters act as screen to divide each group of seats so as to provide a private space for customers. The reception table is decorated with plain color and texture faced wall in hollow wooden pattern.

Tea House

On the wall of the tea house hangs Lijiang local rice-paper painting, describing the long history of tea. Furniture mix-matches Chinese style, Dongba style, Naxi ethnic style and modern style to meet the refined interest of tea-tasting in the landscape of the Jade Dragon Snow Mountain.

藏文化符号 高原奢华情调

拉萨圣地天堂洲际大饭店

Tibetan Cultural Symbol　Plateau Luxury Flavor

InterContinental Lhasa Paradise, Lhasa

酒店业主：会展旅游集团　室内设计：北京圣唐古驿设计事务所
项目地址：西藏拉萨市江苏大道1号　项目面积：约220 000平方米　采编：陈惠慧

Ownership: Exhibition & Travel Group (ETG) China　Interior Design: Beijing Shengtang Guyi Design
Location: No.1 Jiangsu Avenue, Lhasa, Tibet　Site Area: about 220,000 m²　Contributing Coordinator: Chen Huihui

佛教圣地拉萨有着浓郁的宗教氛围，同时也因其位于青藏高原，有着秀美的自然风光，保有一份原生态的气息。拉萨圣地天堂洲际大饭店享有着拉萨河的自然资源，建筑拔地而起，在设计上传承藏族独特的建筑元素。白色的建筑外观与雪域高原相融合，而室内设计在欧式风格下，充分运用藏族元素，如穹顶设计、石材的运用等，突显出酒店奢华的同时又传达出生态、粗犷的野性气质。

Lhasa is a Buddhist holy land with a thick religious atmosphere. Because Lhasa situates in the Qinghai-Tibet Plateau, it enjoys beautiful natural landscapes and a primitive breath. The InterContinental Lhasa Paradise is standing by the Lhasa River, so it owns the river's treasured natural resources. The design inherits Tibetan characteristic elements. The white architectural appearance blends well with the snow-covered plateau. Interior design fully adopts Tibetan symbols on the basis of European style. The dome and stone material applications not only highlight the luxury of the hotel, but also express an ecological and wild flavor.

IHG
INTERCONTINENTAL
INTERCONTINENTAL

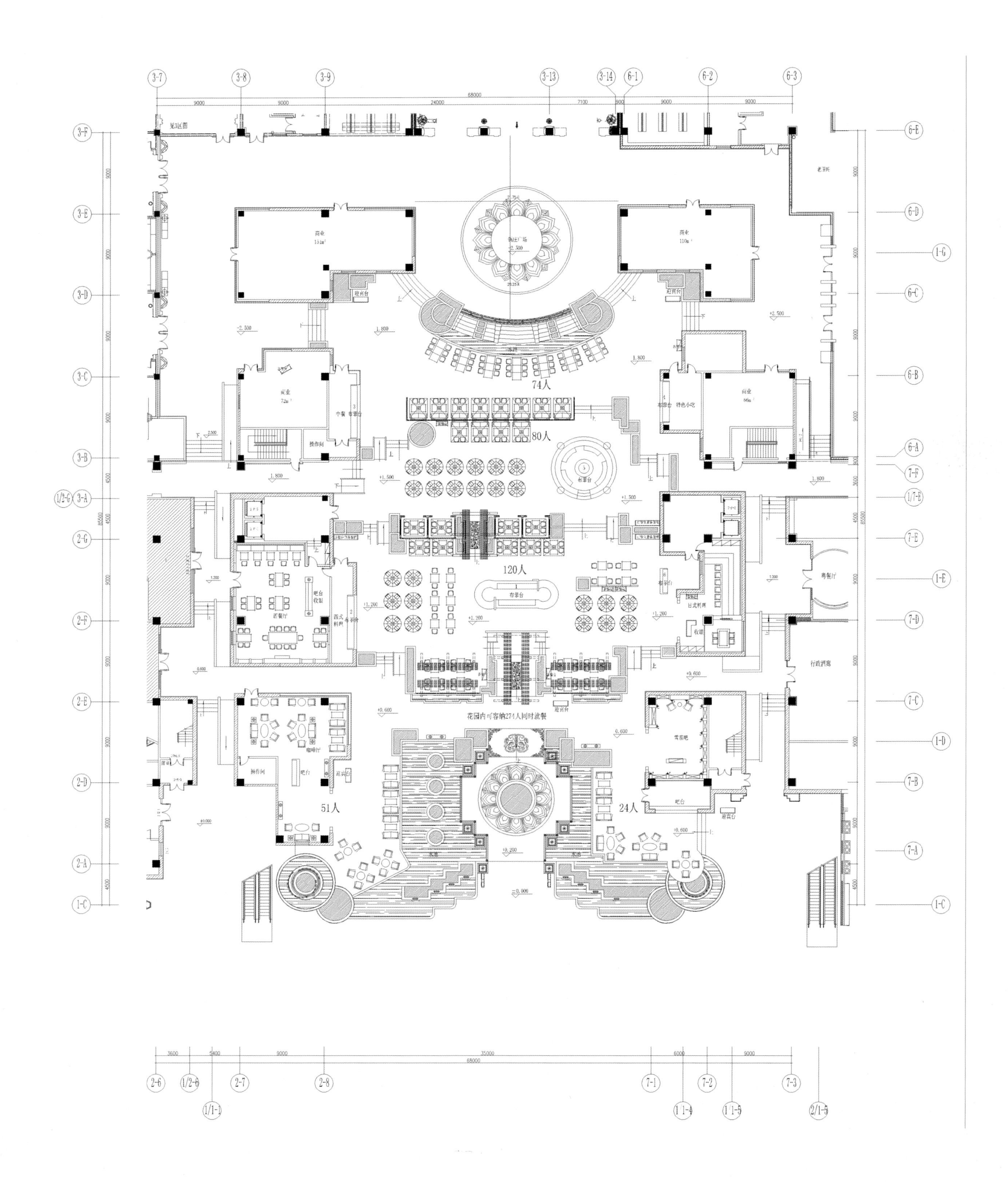

一层花园总平面布置图 1F Garden Master Plan

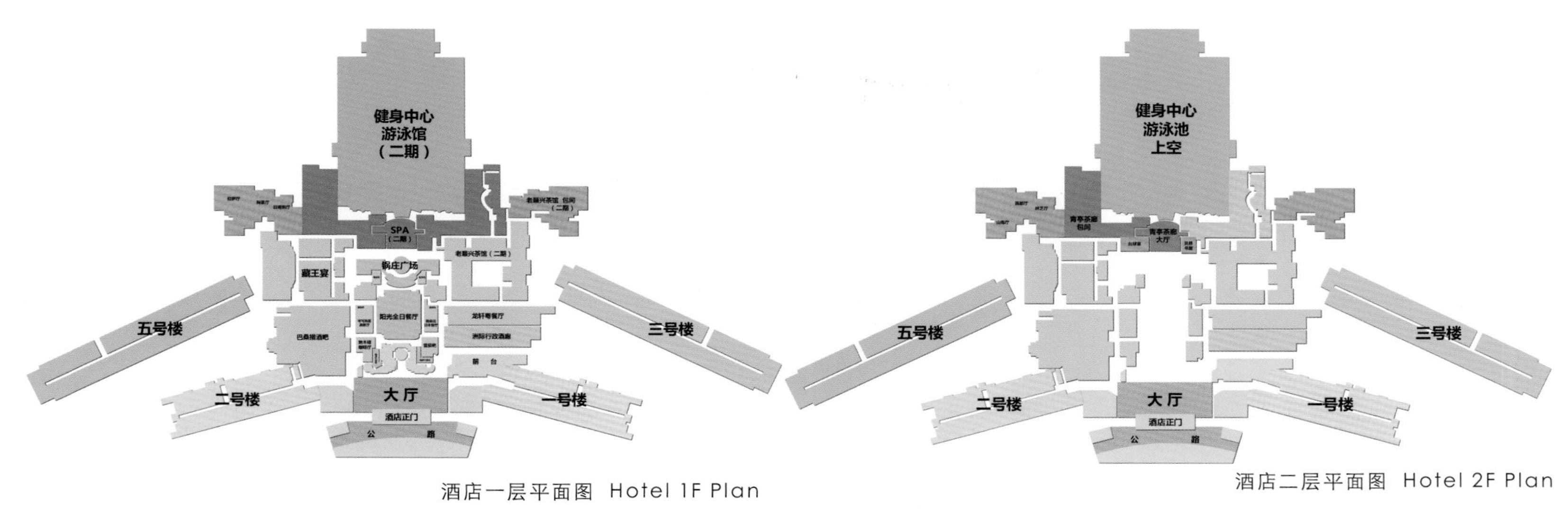

酒店一层平面图 Hotel 1F Plan

酒店二层平面图 Hotel 2F Plan

区位分析

项目位于风景优美的拉萨河畔，距离西藏地标性建筑布达拉宫仅几公里之遥，同拉萨市政府近在咫尺。

项目背景

拉萨位于西藏高原的中部、喜马拉雅山脉北侧，海拔3 650米，是中国西藏自治区的首府，西藏的政治、经济、文化和宗教中心，是中国历史文化名城，也是藏传佛教圣地，有着秀美的自然风光，富有浓郁的民族气息和浓厚的宗教色彩，其建筑以民族风格为主。白色的建筑外观与雪域的圣洁相呼应，亦将建筑与自然和谐相融。

业态功能

项目总体规划房间2 000间，一期开设拥有安装中央供氧设施的豪华客房及套房472间，10个不同风味的餐厅及酒吧以及台球室、书吧等豪华设施，二期开设861间客房以及SPA、室内游泳池。 此外还设有会议室、茶艺室等配套功能空间。

Location Overview

The restaurant sits by the beautiful Lhasa River, just miles away from the Tibet landmark-the Potala Palace, and the Lhasa Municipal Government is nearby.

Project Context

Lhasa is in the center of the Tibetan Plateau and the north of the Himalayas, at an altitude of 3,650 m. It is the capital city of the Tibet Autonomous Region, and meanwhile it is also the center of politics, economy, culture and religion in Tibet, a historic and cultural city and a Buddhist paradise. Lhasa enjoys beautiful natural resources, and because it contains rich ethnic ambiance and profound religious tone, its architecture is in ethnic style.

Service Functions

The overall planning include 2,000 rooms. The first phase provides 472 deluxe guest rooms and suites with central oxygen supply facilities, 10 different flavor dining halls and other luxury facilities, for instance billiard rooms and boor bars; the second phase includes 861 guest rooms, SPA centers, swimming pools, and supporting functional spaces, such as conference rooms and tea houses.

建筑设计

项目建筑外观在体现藏区“尚白”的基础上，把“漂浮的白云，层叠的雪山”融入建筑设计中，体现了雪域高原的特色。

Architectural Design

The appearance of the architecture is basing on white that the color is highly respected by Tibetans. Floating white clouds and layered snow mountains are blended in the building to present characteristics of snow-covered plateau.

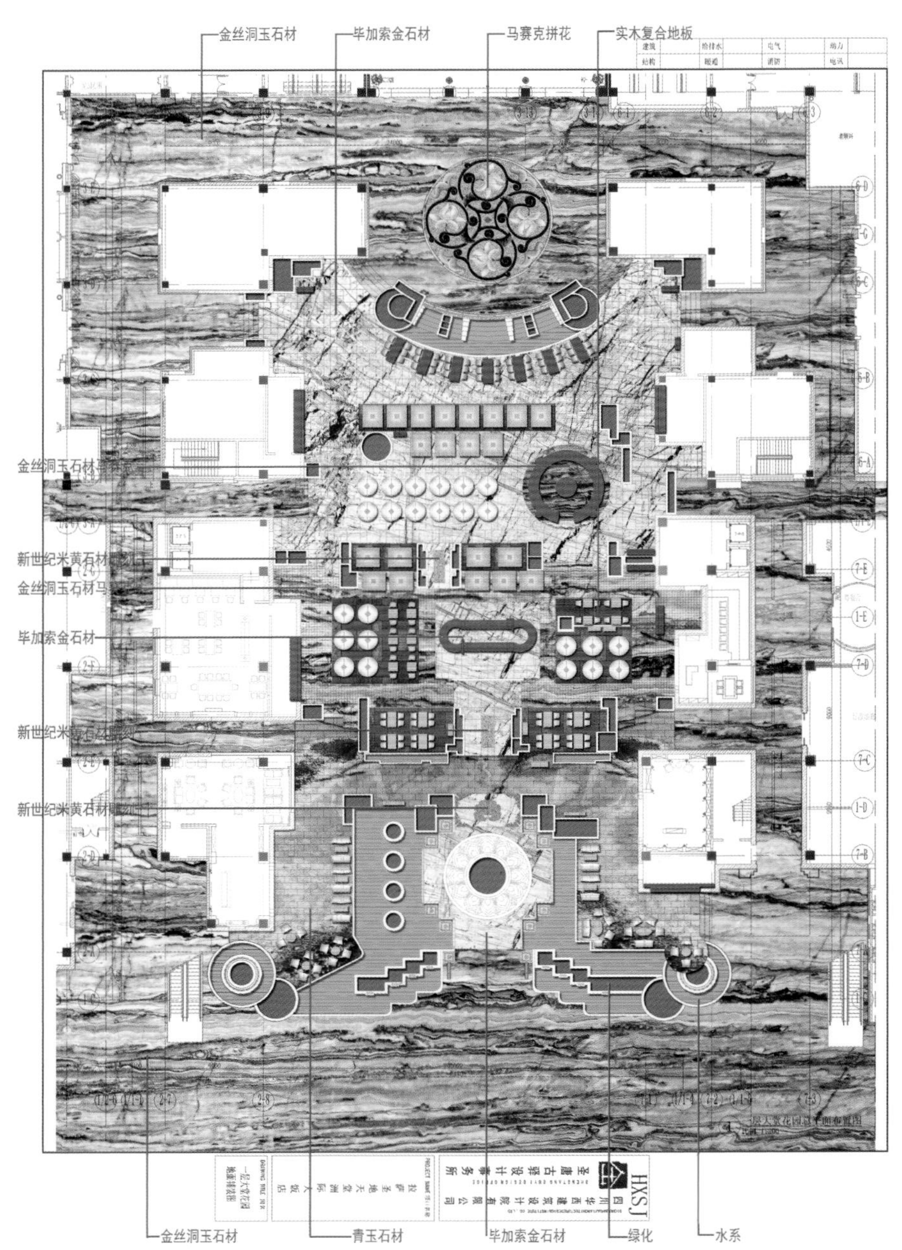

室内设计

室内设计秉承国际标准与西藏地域文化和谐共生的原则展开设计。在风格上，室内设计以欧式风格为基础，进行空间的处理，并将能够代表西藏文化的符号提取出来，使其巧妙地融入其中。

大堂

大堂以欧式古典风格为设计基调，凸显奢华气派。在空间分隔及处理上，继承了欧式风格中豪华、多变的视觉效果，也吸取了藏式风格中唯美、律动的细节元素，力求空间层次的丰富化、氛围的文化感。

花园式全日餐厅、大堂前厅

酒店公共区域设计延续了圣地天堂大酒店建筑外观天然、粗旷的雪域高原风情。

酒店大堂将室内即室外的建筑理念运用得淋漓尽致，宏大的外罩是人工的天穹，天穹中一座座藏式建筑高高耸立，此种空间构造超乎想象，突显了整个空间的自由感。大堂内建筑外墙的石材运用充分体现了藏族建筑特有的文化符号，突显质朴大方、狂野奔放的气质。

Interior Design

The interior design abides by a coexisting principle of international norm and Tibetan culture. It applies European style as basis to organize spaces, and meanwhile some typical Tibetan cultural symbols are abstracted to blend in.

Lobby

The lobby design adopts European classical style to highlight luxury. As for the space-partitioning, the design inherits European sumptuous and versatile visual effects and blends in some Tibetan beautiful and energetic elements so as to present a space in multiple and cultural ambiance.

Garden-like Dining Hall and Lobby Antechamber

The public area of the hotel follows the same appearance of natural and wild snow plateau flavor.

The lobby hall fully utilizes indoor and outdoor architectural concepts. The grand encloser is a got-up dome where a Tibetan building towers on to create a space beyond imagination, bringing a liberty feeling. Tibetan architecture cultural symbols are imprinted on stone exterior walls, which brings a pristine and wild flair.

行政酒廊

行政走廊整体色调以深咖啡色、木色、乳白、金色为主；家居风格现代时尚，充满设计感。装饰精致的艺术品、现代油画等令整个空间低调中透着不羁，深沉而华贵。

二层休息区、娱乐区

休息区以宫廷建筑为代表的中国古典建筑的室内装饰设计艺术风格，气势恢宏、壮丽华贵、高空间、大进深、雕梁画栋、造型讲究对称、色彩讲究对比，装饰材料以木材为主。设计融合更多的后现代手法，把传统的结构形式通过重新设计组合以藏传民族特色的标志符号出现，呈现庄重与优雅双重气质。

茶室

茶室使用大量的木料，以木料的质感古朴，营造出一种亲切的品茗氛围。软装部分有很多藏族文化元素的修饰，特别是屋顶。设计在细节中穿插着藏文化的符号，色调以闲淡雅致为主，更加体现高雅尊贵的气质。

客房

客房设计采用独具古典风格的设计理念，与当地高原景观自然融合。

Executive Lounge

The whole executive lounge is dominant by dark coffee, natural wood color, cream and golden colors. Furniture is full of design feeling of modern fashion. Exquisite artworks and modernist oil paintings create an uninhibited, deep and sumptuous ambiance in the low key space.

Resting and Recreational Area on Second Floor

The resting area adopts Chinese classical palace interior decoration style. It is an imposing and magnificent area with high space and long depth. Carved beams and painted rafters stand symmetrically, and meanwhile they stress on color comparisons with rich wooden ornaments. The overall design blends in post-modern methods to reconstruct traditional architectural configurations so as to highlight Tibetan characteristic symbols, solemn and elegant.

Tea House

The tea house applies a large number of wooden materials, because wood is conducive to create a primitive and intimate tea tasting atmosphere. Soft-mounted parts are embellished by Tibetan cultural elements, especially on the roof, and detail design adds up Tibetan cultural symbols as well. The light and elegant colors further express a noble flair.

Guest Room

The guest room decoration adopts classic style to echo local natural plateau landscapes.

索引

Index